Aberration-Corrected Analytical Transmission Electron Microscopy

Current and future titles in the RMS-Wiley Imprint

Published

Principles and Practice of Variable Pressure/Environmental Scanning Electron Microscopy (VP-ESEM) – Debbie Stokes

Aberration-Corrected Analytical Electron Microscopy – Edited by Rik Brydson

Forthcoming

Understanding Practical Light Microscopy – Jeremy Sanderson

Atlas of Images and Spectra for Electron Microscopists – Edited by Ursel Bangert

Diagnostic Electron Microscopy – A Practical Guide to Tissue Preparation and Interpretation – Edited by John Stirling, Alan Curry & Brian Eyden

Practical Atomic Force Microscopy – Edited by Charles Clifford

Low Voltage Electron Microscopy – Principles and Applications – Edited by Natasha Erdman & David Bell

Focused Ion Beam Instrumentation: Techniques and Applications – Dudley Finch & Alexander Buxbaum

Aberration-Corrected Analytical Transmission Electron Microscopy

Edited by

Rik Brydson

Published in association with the Royal Microscopical Society

Series Editor: Susan Brooks

A John Wiley & Sons, Ltd., Publication

Library of Congress Cataloging-in-Publication Data

Aberration-corrected analytical transmission electron microscopy / edited by Rik Brydson ... [et al.].
    p. cm.
   Includes bibliographical references and index.
   ISBN 978-0-470-51851-9 (hardback)
  1. Transmission electron microscopy. 2. Aberration. 3. Achromatism. I. Brydson, Rik.
   QH212.T7A24 2011
   502.8′25–dc23

                                                                              2011019731

A catalogue record for this book is available from the British Library.

Print ISBN: 978-0470-518519 (H/B)
ePDF ISBN 978-1119-978855
oBook ISBN: 978-111-9978848
ePub ISBN: 978-1119-979906
Mobi ISBN: 978-1119-979913

Set in 10.5/13pt Sabon by Laserwords Private Limited, Chennai, India

Contents

List of Contributors

Andrew Bleloch, School of Engineering, University of Liverpool, Liverpool, UK

Lawrence Michael Brown FRS, Robinson College, Cambridge and Cavendish Laboratory, Cambridge, UK

Rik Brydson, Leeds Electron Microscopy and Spectroscopy (LEMAS) Centre, Institute for Materials Research, SPEME, University of Leeds, Leeds, UK

Alan J. Craven, Department of Physics and Astronomy, University of Glasgow, Glasgow, Scotland, UK

Peter J. Goodhew FREng, School of Engineering, University of Liverpool, Liverpool, UK

Sarah J. Haigh, Department of Materials, University of Oxford, Oxford, UK; University of Manchester, Materials Science Centre, Manchester, UK

Nicole Hondow, Leeds Electron Microscopy and Spectroscopy (LEMAS) Centre, Institute for Materials Research, SPEME, University of Leeds, Leeds, UK

Angus I. Kirkland, Department of Materials, University of Oxford, Oxford, UK

Peter D. Nellist, Department of Materials, Corpus Christi College, University of Oxford, Oxford, UK

Quentin Ramasse, SuperSTEM Facility, STFC Daresbury Laboratories, Daresbury, Cheshire, UK

Mervyn D. Shannon, SuperSTEM Facility, STFC Daresbury Laboratories, Daresbury, Cheshire, UK

Gordon J. Tatlock, School of Engineering, University of Liverpool, Liverpool, UK

All contributors helped in the preparation and editing of this book.

Preface

Electron Microscopy, very much the imaging tool of the 20th Century, has undergone a steep change in recent years due to the practical implementation of schemes which can diagnose and correct for the imperfections (aberrations) in both probe-forming and image-forming electron lenses. This book aims to review this exciting new area of 21st Century analytical science which can now allow true imaging and chemical analysis at the scale of single atoms.

The book is concerned with the theory, background and practical use of transmission electron microscopes with lens correctors which can mitigate for, and to some extent control the effects of spherical aberration inherent in round electromagnetic lenses. When fitted with probe correctors, such machines can achieve the formation of sub-Angstrom electron probes for the purposes of (scanned) imaging and also importantly chemical analysis of thin solid specimens at true atomic resolution. As a result, this book aims to concentrate on the subject primarily from the viewpoint of scanning transmission electron microscopy (STEM) which involves correcting focused electron probes, but it also includes a comparison with aberration correction in the conventional transmission electron microscope (CTEM) where the principal use of correctors is to correct aberrations in imaging lenses used with parallel beam illumination.

The book has arisen from the formation in 2001 of the world's first aberration corrected Scanning Transmission Electron Microscope Facility (SuperSTEM http://www.superstem.com) at Daresbury Laboratories in Cheshire in the UK. This originally involved a consortium of the Universities of Cambridge, Liverpool, Glasgow and Leeds, the Electron Microscopy and Analysis Group of the Institute of Physics and the Royal Microscopical Society and, very importantly, funded by the Engineering and Physical Sciences Research Council (EPSRC). Following its inception we have organised four postgraduate summer schools in 2004,

2006, 2008 and 2010. The current text has evolved from these Summer Schools and aims to be a (detailed) handbook which although introductory, has sections which go into some depth and contain pointers to seminal work in the (predominantly journal) literature in this area. We envisage that the text will be of benefit for postgraduate researchers who wish to understand the results from or wish to use directly these machines which are now key tools in nanomaterials research. The book complements the more detailed text edited by Peter Hawkes (*Aberration corrected Electron Microscopy*, Advances in Imaging and Electron Physics, Volume 153, 2008).

The form of the handbook is as follows:

In Chapters 1 and 2, Peter Goodhew and Gordon Tatlock introduce general concepts in transmission electron microscopy and electron optics. In Chapter 3 Mick Brown details the development of the concept of the scanning transmission electron microscope which arose from the pioneering vision of Albert Victor Crewe, who was notably born in Bradford in the West Riding of Yorkshire and was a graduate of Liverpool University. Probe forming lens aberrations and their diagnosis and correction are investigated further by Andrew Bleloch and Quentin Ramasse in Chapter 4. The theory of STEM imaging is outlined by Pete Nellist in Chapter 5, whilst the detailed instrumentation associated with STEM is given by Alan Craven in Chapter 6. Analytical spectroscopy in STEM and the implications of STEM probe correction are introduced in Chapter 7 by Rik Brydson and Nicole Hondow. Mervyn Shannon reviews some examples and applications of aberration corrected STEM in Chapter 8. Finally, in Chapter 9, Sarah Haigh and Angus Kirkland make comparisons with image correction in CTEM. All authors have very considerable experience in transmission electron microscopy and also aberration correction from either a practical or applied perspective and we have attempted to integrate the various separate chapters together so as to form a coherent text with a common nomenclature detailed in Appendix B. Although I have taken the nominal lead in editing the text, it has been a joint effort by all authors.

Tremendous thanks must go to all those associated with SuperSTEM over the years notably Meiken and Uwe Falke, Mhari Gass, Kasim Sader, Bernhard and Miroslava Schaffer, Budhika Mendis, Ian Godfrey, Peter Shields, Will Costello, Andy Calder, Quentin Ramasse, Michael Saharan, Dorothea Muecke-Herzberg, Marg Robinshaw, Ann Beckerlegge and Uschi Bangert. Dedicated SuperSTEM students have been invaluable and have included Sarah Pan, Paul Robb, Peng Wang, Linshu Jiang, Dinesh Ram, Sunheel Motru and Gareth Vaughan.

The final statements concerning the book should belong to Charles Lutwidge Dodgson (aka Lewis Carroll) who was born at Daresbury Parsonage less than a mile from the SuperSTEM laboratory and who famously wrote, '*Through the* (aberration-corrected) *Looking Glass*' and '*Alice in Wonderland*'

'*Begin at the beginning and go on till you come to the end: then stop.*'

But. . . . . . .

'*I don't believe there's an atom of meaning in it.*'

And. . . . .

'*What is the use of a book, without pictures or conversations?*'

Rik Brydson, Leeds 2011.

1

General Introduction to Transmission Electron Microscopy (TEM)

Peter Goodhew

School of Engineering, University of Liverpool, Liverpool, UK

1.1 WHAT TEM OFFERS

Transmission electron microscopy is used to reveal sub-micrometre, internal fine structure in solids. Materials scientists tend to call this *microstructure* while bioscientists usually prefer the term *ultrastructure*. The amount and scale of the information which can be extracted by TEM depends critically on four parameters; the resolving power of the microscope (usually smaller than 0.3 nm); the energy spread of the electron beam (often several eV); the thickness of the specimen (almost always significantly less than 1 μm), and; the composition and stability of the specimen. The first and second of these depend largely on the depth of your pocket – the more you spend, the better the microscope parameters. The third is usually determined by your experimental skill while the last depends on luck or your choice of experimental system. The slim book by Goodhew *et al.* (Goodhew, Humphreys and Beanland,

Aberration-Corrected Analytical Transmission Electron Microscopy, First Edition.
Edited by Rik Brydson.
© 2011 John Wiley & Sons, Ltd. Published 2011 by John Wiley & Sons, Ltd.

2001) gives a simple introduction to all types of electron microscopy, whilst the comprehensive text by Williams and Carter provides more detailed information on transmission electron microscopy (Williams and Carter, 2009).

The two available types of TEM – CTEM and STEM – differ principally in the way they address the specimen. The conventional TEM (CTEM) is a wide-beam technique, in which a close-to-parallel electron beam floods the whole area of interest and the image, formed by an imaging (objective) lens after the thin specimen from perhaps 10^6 pixels, is collected in *parallel*. The scanning TEM (STEM) deploys a fine focused beam, formed by a probe forming lens before the thin specimen, to address each pixel in *series* as the probe is scanned across the specimen. Figure 1.1 summarises these differences. In both types of instrument analytical information from a small region is usually collected using a focused beam. The smallest region from which an analysis can be collected is defined by the diameter of this beam and hence the corresponding through-thickness volume in the specimen within which various inelastic scattering (energy-loss) processes take place.

As we will discuss in Chapter 2, the image resolution in CTEM is primarily determined by the imperfections or aberrations in the objective lens, whilst in a STEM instrument the resolution of the scanned image (as well as the analytical resolution described above) is determined largely by the beam diameter generated by the probe-forming lens which is also

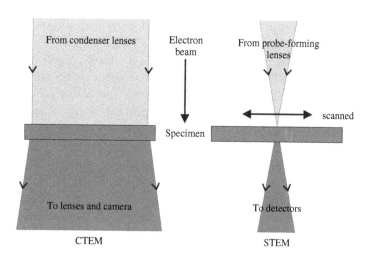

Figure 1.1 The electron beam in CTEM and STEM instruments. Note that in many STEM instruments the beam actually travels upwards rather than down as shown here.

limited by aberrations. In recent years, technical difficulties have been overcome principally due to increases in computing power which have allowed both the diagnosis and correction of an increasing number of these lens aberrations (Hawkes, 2008). Aberration correction could in principle be applied to any magnetic lens in any microscope. However there are in practice two key areas where it is employed (in some cases in tandem): the condenser/illumination system and the objective lens. In STEM, of course, there is only a condenser system, which is better called a probe-forming lens (confusingly in a STEM this is often referred to as an objective lens – however, the important point is that it lies before the specimen) – making full aberration correction significantly cheaper! One of the main benefits of the correction of spherical aberration in STEM, is in the reduction of the 'beam tails' so that a fine beam positioned on a specified column of atoms does not 'spill' significant electron intensity into neighbouring columns. As we will see, this has big implications for the STEM-based techniques of high angle annular dark field (HAADF) or 'Z contrast' imaging and electron energy loss spectrometry (EELS) analysis (see section 1.3).

Although CTEM instruments far outnumber STEMs the advantages of the STEM approach – mainly the superior control over the positioning of the beam and better scope for detecting useful signals – are widely recognised and many CTEM instruments now include some STEM capabilities. At the present state of the art, aberration corrected STEM instruments probably offer the best (i.e. most sensitive and highest spatial resolution) analytical capabilities. Therefore, the primary focus of this handbook will be aberration correction in STEM, and Chapters 3–6 introduce and describe this approach in more detail. However aberration correction in CTEM is discussed for comparative purposes in Chapter 9.

All the potential ways in which imaging or analytical information can be extracted by any type of TEM depend critically on the interactions between the electrons in the beam and the electrons and/or nuclei of the atoms which comprise the specimen. We therefore consider some of the most useful interactions in the next section. As this text is, by design, somewhat brief, the description of these interactions is principally an introduction and further detail may be found in the extremely comprehensive text of Egerton (Egerton, 1996).

1.2 ELECTRON SCATTERING

Electron scattering within the specimen can be characterised in various ways. In what follows the *primary electron* is a single high energy electron in the beam which (for TEM specimens of useful thickness) passes through the specimen, whilst a *secondary electron* is created within the specimen and might or might not subsequently leave the specimen.

Simplistically we refer to *elastic scattering*, implying that the energy lost by the primary electron is too small to be detected, or *inelastic scattering*, implying an energy loss big enough to be detected in the microscope. Staying at the simplistic level, we can say that elastic scattering occurs mainly from interactions with atomic nuclei (or the whole electrostatic field of the atom), is responsible for diffraction from crystalline specimens and is used to reveal structure; meanwhile inelastic scattering mainly involves electron-electron interactions and is mainly exploited for analysis – for example by EELS or energy dispersive X-ray (EDX) analysis, techniques which are discussed in Chapter 7.

A second way of characterising electron scattering is in terms of the process, that is what exactly happens or what is produced. Thus we can distinguish Rutherford or Compton scattering, plasmon or phonon production, ionization and several other processes (Egerton, 1996).

A third way of describing scattering within a specimen is to indicate the average number of scattering events of a specific type during the transit of a single primary electron. In this context *single scattering* implies that only one or two (or zero) events of the specified type are likely to occur during the passage of a single primary electron. *Plural scattering* then implies that a countable number of events occur, that is a sufficiently small number that Monte Carlo calculations (these are computer simulations based on probabilities of occurrence) might be able to keep track of them. Finally *multiple scattering* implies that the scattering events are so frequent that an almost uncountable number occur during the lifetime of the primary electron within the specimen. This is rare in a thin TEM specimen but is of great relevance to scanning electron microscopy (SEM) of (bulk) solids, when at some stage the primary electron loses most of its energy and becomes indistinguishable from any other electron in the specimen.

An important concept related to the frequency with which any particular interaction occurs is the *mean free path*, Λ, which is the average distance travelled by the primary electron between interactions. The same idea can be alternatively expressed by the *cross-section*, σ, the apparent cross-sectional area offered to the primary electron by the scattering entity. These two quantities are simply related via:

$$\Lambda = 1/(N_V \sigma) \tag{1.1}$$

where N_V is the number of scatterers (atoms) per unit volume. The appropriate cross-section depends on the solid angular range, Ω, over which the scattering is detected, so scattering probability is often referred to in terms of the differential cross-section $d\sigma/d\Omega$.

Electron interactions are sometimes characterised as spatially *coherent* (resulting in an in-phase effect such as diffraction from neighbouring scattering centres) or *incoherent* (not in phase, resulting in uncorrelated events from different scattering centres, such as high angle elastic scattering or the whole range of inelastic scattering events). It is important, however, to realise that in electron microscopy there is only one electron in the microscope at a time – at the currents commonly used, individual electrons are spaced perhaps a metre apart. The concept of coherence is therefore not easy to comprehend, and one hears statements of the type 'each electron can only be coherent with, and thus interfere with, itself.' Coherent electrons each 'make the same pattern'. In some circumstances it is also necessary to worry about *temporal* coherence, as well as *spatial* coherence. This is related to the variation in wavelength among the electrons and thus to the energy stability of the electron gun. Again it is conceptually difficult: coherence lengths are inversely proportional to the energy spread of the beam probe (which in STEM we usually try to keep below 1 eV) and they have magnitudes of the order of hundreds of nm. The conceptually difficult bit is that we are concerned about the temporal coherence, characterised by a length less than a μm, of electrons which are a metre or so apart and therefore at first sight independent and unaware of each other.

Finally, for the moment, we need to define the wave vector, $\mathbf{k}$, and scattering vector, $\mathbf{q}$. We will use the convention that $|\mathbf{k}| = 2\pi/\lambda$, rather than $1/\lambda$. If $\mathbf{k}$ is the incident wave vector and $\mathbf{k'}$ is the resultant scattered wave vector then the scattering vector is simply the difference between the two, $\mathbf{q} = \mathbf{k} - \mathbf{k'}$, as shown in Figure 1.2. Note here the direction of the scattering vector $\mathbf{q}$ is taken as that of momentum transfer *to* the specimen (equal to $h/2\pi$ times $\mathbf{q}$) which is opposite to the direction of the wavevector change of the fast electron.

In order to understand the range of potential analytical techniques available to the electron microscopist we need to understand a little about each of the more common (or more useful) interactions. To aid the subsequent description, Figure 1.3 displays a schematic spectrum of electron energies after their transmission through a thin specimen (in this case an EELS spectrum of calcium carbonate).

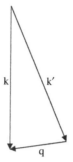

Figure 1.2 Vector diagram of the incident and scattered wavevectors, **k** and **k′**. The corresponding scattering vector is **q**.

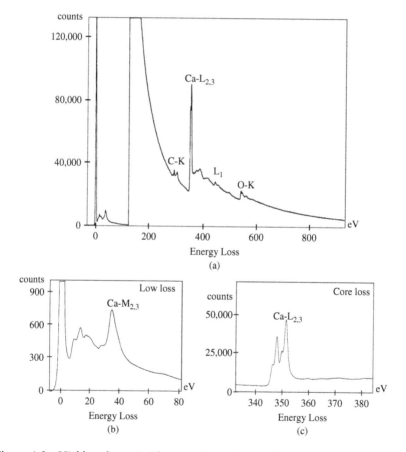

Figure 1.3 Highly schematic Electron Energy Loss (EELS) spectrum of calcium carbonate showing the zero loss peak, a band gap, interband transitions, a plasmon peak (followed by a change in intensity scale) and ionisation edges: a carbon K-edge at ca. 285 eV, a calcium $L_{2,3}$ edge at ca. 350 eV and an oxygen K-edge at ca. 532 eV. Adapted from the *EELS Atlas* by C.C. Ahn and O.L. Krivanek.

1.2.1 Elastic Scattering

Elastic scattering involves no (detectable) energy loss and is represented by the peak at zero energy loss in Figure 1.3. It is usually described in terms of Rutherford scattering from an atom or ion. Introductory texts give the probability of scattering through an angle θ as $p(\theta)$ α $1/E_0^2\sin^4(\theta/2)$, where E_0 is the electron (kinetic) energy. This expression implies strongly forward-peaked scattering in which the scattering is very much more likely to be through a very small angle. However, Rutherford scattering ignores screening of the nucleus by localised electrons and overestimates scattering at low angles. It also gives an infinite cross section if integrated over all angles! A better treatment leads to (Egerton, 1996):

$$d\sigma/d\Omega = 4\gamma^2 Z^2/a_0^2 q^4 \quad \text{where } q = |\mathbf{q}| = 2k\sin(\theta/2) \quad (1.2)$$

Here γ (where $\gamma = [1 - v^2/c^2]^{-1/2}$) is the relativistic factor for the electron velocity (v) relative to the speed of light (c), Z is the atomic number, a_0 is the Bohr radius, q is the magnitude of the scattering vector and $k = |\mathbf{k}|$ is the magnitude of the wave vector of the incident electron ($k = 2\pi/\lambda$ for an electron of wavelength λ, as defined above). Screening can be taken care of by replacing q^2 by $q^2 + r_0^{-2}$ where r_0 is a screening radius. Here it is still the case that, at small angles, $d\sigma/d\Omega$ α $1/E_0^2(\theta/2)^4$ but now the magnitude of scattering is smaller and more realistic.

Low angle elastic scattering, because of its spatially coherent nature, can give information on the relative positions of atoms within the sample and is heavily exploited in diffraction techniques, both selected area and convergent beam, and (together with inelastic scattering) in the formation of Kikuchi bands. However, apart from HAADF imaging (see section 1.3 below) and some recent quantitative CBED developments, it is rare that elastic scattering intensities are measured or interpreted in electron microscopy, so inaccuracies in the Rutherford-based approach of the previous paragraph are usually not important.

An important point of terminology in terms of elastic scattering is the distinction between the kinematical and the dynamical scattering regimes. These terms refer to the approximations which it is appropriate to use. In a kinematical scattering regime each particle in the illuminating probe has only a small chance of being scattered, and the approximation can be made that the incoming probe loses negligible intensity. This is a reasonable approximation in many cases of X-ray and neutron scattering but is usually only a poor approximation for electrons passing through a crystal owing to their large interaction with matter. In contrast, in the dynamical regime multiple scattering occurs, with multiple beams excited within the specimen which can then interfere. Intensity which oscillates with depth is then the norm for any beam.

Finally, an important concept in STEM is electron channelling, which is explored further in section 5.4. When the electron probe is located over an atomic column in a crystal, the electrons tend to be trapped by the attractive, periodic potential of the atoms. The process acts somewhat like a waveguide, and owing to the dynamical nature of the process, oscillations of the intensity of the wave on the atom sites with depth in the crystal are seen.

1.2.2 Inelastic Scattering

Most electron spectrometry associated with TEM involves measuring the energy of the primary electrons after they have passed through the specimen as displayed in Figure 1.3. Most such electrons have only lost less than a few tens of eV so are still highly energetic, typically around 100 keV or 200 keV. The spectrometers available to electron microscopists have a useful energy resolution limited to 0.1 eV or more in these circumstances, and the energy spread in the probe is likely to be at least 0.25 eV, so microscopists cannot easily avail themselves of processes with energy loss less than 1 eV or so. Thus, despite their numerical preponderance (arising from their short mean free path), inelastic scattering events leading to the production of *phonons* are not usefully exploited and are subsumed in the zero loss peak. Phonon scattering contributes to the background of images and diffraction patterns, and is ultimately responsible for heating the specimen, but because little can be deduced from its intensity or angular distribution it is rarely useful (at least at present). An effect which is in practice indistinguishable from phonon excitation (at least at room temperature) is thermal diffuse scattering (TDS), which is in effect the interaction of the probe electrons with pre-existing phonons. Both effects contribute to the background of many images, diffraction patterns and spectra, but the two effects cannot be separated unless the specimen is cooled, when TDS will be reduced while phonon excitation will not.

1.2.2.1 Plasmon Scattering

A plasmon is a collective oscillation of 'free' electrons induced by the transmission incident electron. As such it is rather a non-localised scattering event occurring over a number of atomic sites. It dies away very quickly but dissipates significant energy, usually in the range up to 30 eV (Figure 1.3). Plasmon excitation can therefore be responsible for energy losses of tens of eV from a primary electron. The mean free path depends on the specimen but is typically in the 100 nm range. In specimens which are too thick for the most accurate or precise analysis, two or more plasmons can be excited by a single primary electron, leading to 'double' and 'triple' plasmon peaks in the EELS spectrum (section 7.4).

Plasmon-like oscillations occur even in non-metallic materials without apparent free electrons. Oscillations with different characteristic energies from bulk plasmons are also associated with surfaces and interfaces. All these phenomena are loosely referred to as plasmons. In the spectrum in Figure 1.3, it is apparent that plasmons generally overlap in energy with signals due to the transitions of single electrons between the outermost bands of electron energies. This combined valence electron scattering also contributes to the background of the EELS spectrum at higher energies which has considerable intensity at higher scattering angles.

Plasmon peaks are generally broad, with energies which are not uniquely representative of the specimen (unlike characteristic X-rays or Auger electrons, for instance). Egerton (1996) gives a list of measured and calculated values for many materials. They are a major and intense feature of EELS spectra, but since early attempts in the 1960s have only rarely been used for analysis, although recently small shifts in plasmon peak energies have been used to detect changes in composition and there is increasing interest in the excitation of plasmon modes in nanostructures.

1.2.2.2 Inner Shell Ionisation

The most useful, but not most frequently occurring, interaction is the ejection by the primary electron of a localised bound electron from a sample atom. The cross-section for the excitation of an inner shell electron is usually small (mean free paths are often >1 μm) but the energy loss is often quite sharp and very characteristic of both the element and its state of bonding. Inner shell excitation is thus at the heart of energy loss spectrometry in the EM, leading to small but easily-recognised 'edges' in EEL spectra (Figure 1.3).

Following ionisation, the atoms of the specimen will relax, emitting either a characteristic X-ray (superimposed on a set of continuous background X-rays known as *Bremsstrahlung*) or an Auger electron.

X-rays are relatively easily detected in the column of a TEM and X-ray emission forms the basis of EDX analysis (see section 1.3). Auger electrons, on the other hand, require quite large spectrometers and because they are charged need sophisticated techniques to extract them from the microscope whilst preserving information about their energy. They are therefore rarely used for analysis in the TEM.

Inner shell excitations are dominated by the structure of the single atom, so are capable of being treated theoretically using simple models. Egerton (1996) and also Brydson (2001) give substantial detail of appropriate theories, the prediction of cross-sections, and the appearance of edges in an experimental spectrum. It is worth noting that significant information about the density of empty states and thus the bonding within the specimen can be deduced from near-edge fine structure (ELNES, at the start of the edge) and extended fine structure (EXELFS, after the main edge).

All these topics are examined in more detail in Chapter 7.

1.3 SIGNALS WHICH COULD BE COLLECTED

A 'standard' analytical CTEM is able to collect bright field and dark field images, selected area and convergent beam diffraction patterns, characteristic X-ray spectra (EDX) and electron energy loss (EEL) spectra. For an introduction to all of these, consult Goodhew et al. (2001). Here, we primarily focus on STEM.

A 'standard' STEM offers a similar set of possibilities, but differently optimised and set up for mapping. In addition it offers annular dark field (ADF) imaging. This is discussed further in Chapter 3, section 3.4. Bright field STEM images contain all the standard contrast mechanisms as experienced in bright field CTEM: mass-thickness, diffraction and phase contrast (Goodhew et al., 2001). Of particular interest to users of aberration-corrected instruments is high-angle annular dark field (HAADF) imaging. High-angle in this context signifies several tens of mrad, which is beyond the angle at which diffraction maxima (spots) of any significant intensity are found. High-angle scattered electrons are few in number, and mostly result from Rutherford scattering. Their great advantage for imaging and analysis is that they are usually insensitive to structure and orientation but strongly dependant on atomic number, with the intensity varying as Z^ζ where ζ lies between 1.5 and 2 and is often quoted for most experimental setups as being around 1.7. A HAADF image, collected from electrons scattered in the angular range perhaps 50 to 200 mrad, therefore has a local intensity which strongly depends on composition, but depends less strongly on structure or precise orientation. If the specimen is uniformly thick in the area of interest the HAADF intensity can be directly related to the average atomic number in the column at each pixel. If the beam is less than one atom dimension in diameter, for instance in an aberration-corrected STEM, then atom column compositional resolution is therefore possible (strictly if we have strong channelling of the probe). This is discussed further in Chapters 5 and 6.

An energy-filtering microscope is one in which the image (or indeed the diffraction pattern) can be formed using electrons whose energy has been selected after they have passed through the specimen. This is achievable in a CTEM with an in-column 'omega filter' or post-column imaging filter (Brydson, 2001) or, more elegantly, in STEM by so-called *spectrum imaging*. In this technique an EELS spectrum is collected at each pixel. The image intensity at each pixel is measured during post processing from a defined energy-loss range within the appropriate spectrum. This effectively produces compositional maps of the specimen, although, as discussed in Chapter 7, care must be taken to select an appropriate energy range and background subtraction routines.

It has been implicitly assumed in this chapter that EELS is the analytical method of choice for high resolution microscopy. There are two key reasons for this dominance; firstly the EELS spectrometer can be located far from the specimen and thus does not interfere with the important region around the specimen, which is already crowded with lens pole pieces and specimen tilting apparatus. Secondly, electrons can be collected by an EELS spectrometer with almost 100% efficiency, so little useful signal is lost.

Despite the advantages of EELS, analysis using characteristic X-rays (EDX) is also widely available. Relaxation of excited atoms by emission of an X-ray becomes more efficient for heavier atoms and EDX might be the analytical technique of choice for specimens containing heavy atoms (say $Z > 30$). However because X-rays cannot be deflected into an appropriate detector their collection will always be inefficient (usually less than 1%) and signal strength will always be a problem from a thin specimen. Both EELS and EDX are more fully discussed in Chapter 7. Other potential, but relatively unexploited techniques applicable to STEM are discussed at the end of section 3.4.

For any signal from a specific feature within a TEM specimen, in order to detect or 'see' it in an image the contrast (strictly defined as $\{S_F - S_B\}/S_B$ where S_F and S_B are the signals from the feature and the surrounding background respectively) needs to be greater (usually between three and five times greater) than the inherent noise level in the background signal, which in Poisson statistics is the square root of the number of counts. Thus detection and visualisation of a feature is highly dependent not only on the resolution but also critically depends on the contrast.

1.4 IMAGE COMPUTING

Most of the easy problems available to the microscopist have already been addressed during those halcyon days when a qualitative argument (sometimes even just hand-waving) was sufficient to explain new features in a micrograph. Many of today's more sophisticated questions can only be answered by a sophisticated analysis of the process of image formation, together with sympathetic processing of experimental images and spectra. Both CTEM and STEM are thus supported by suites of programs designed to manipulate experimental images ('image processing') and to predict what the image of a possible structure would look like in the microscope ('image simulation').

1.4.1 Image Processing

Since image collection (recording) is now almost universally digital, images at any resolution are typically stored as 1024×1024 (for example) datasets, giving about 10^6 image pixels per image. Such images can be manipulated (almost too easily) in a variety of ways. In the future microscopists will no doubt use clever recognition algorithms to locate features of interest and automatically select them for further study, but at present the principal processing techniques (beyond simple control of brightness and contrast) are based around fast Fourier transforms (FFTs) of the image. For images containing a large degree of periodicity (e.g. crystalline materials) such transforms (power spectra corresponding, of course, to diffraction patterns of the same area of the specimen) can be used to filter out 'noise' before back-transformation into a 'better' image. Commercial programs such as Digital Micrograph (Gatan Inc. – www.gatan.com) do this sort of thing, and a lot more, extremely efficiently. However the microscopist must always be aware of the potential for introducing artefacts during what might all too easily be used as 'black box' processing.

Image processing can be used to extract, or sometimes just to make more evident, specifically interesting features of an image. To be applied, it requires little or no *a priori* understanding of the nature of the specimen and typically needs no input parameters.

There is increasingly an ethical issue associated with publishing both analytical and image data from EMs. Although it takes more space (on paper or on-line) we would recommend that the raw data is always published, as well as any processed version of it which might be easier to interpret. Any publication should also make precisely clear what processing has been applied to any image or spectrum. Putting the raw data in the public domain ensures that conclusions can be checked by other researchers.

1.4.2 Image Simulation

Image simulation has been used for many decades and in the 1960s simple programs were in use to predict the appearance of isolated dislocations in thin crystals, with the background assumed to be a diffracting continuum. However the increasing resolving power of modern microscopes has shifted the focus of simulation towards atomic column resolution and structural images. Simulated images can not only help to 'solve' structures but they can also assist the microscopist to distinguish specimen features from instrumental artefacts. They will be referred to later in this book in Chapters 5, 8 and 9.

The simulation of high resolution EM images can only be undertaken, at the present time, by a process in which the microscopist constructs a possible arrangement of atoms and then asks the question 'what would this look like in my microscope under the operating conditions I think I am using?'. Comparison with an actual experimental image is then usually, but not necessarily, performed by eye, to answer the question 'is this what my image shows?'. The further important question 'is this the only structure which might look like this in the microscope?' is, regrettably, not always fully addressed or addressable!

In contrast to image processing, image simulation requires the input of a large number of pieces of information. Programs such as QSTEM (Koch, 2002), TEMSIM (Kirkland, 1998) or JEMS (Stadelmann, 1987), some of which are available as freeware, typically expect the user to define the location and atomic number of every atom in the test structure, the local specimen thickness, the electron energy, the exact value of defocus, the size of apertures, the shape (profile) of the beam (influenced by lens aberrations), atomic scattering factors and the orientation of the specimen. Several of these parameters are very difficult to measure accurately, so simulations are often run for a range of thicknesses and a range of defoci, while the corresponding experimental images are collected as a through-focal series. Despite this complexity many structures have been solved or confirmed by the use of such simulations.

High resolution simulations, for CTEM or STEM, adopt one of two approaches, based on Bloch wave propagation or the multislice principle. Because Bloch wave calculations require greater processing power, and are in some ways less flexible, multislice programs now dominate the field. The principle of a multislice calculation is to cut the trial structure into thin slices perpendicular to the electron beam and compute, for each slice in turn, its effect on the phase of the slowly-varying part of the electron wave function. The wave is then propagated to the next slice. Under appropriate conditions the effects of each slice can be added and the wave propagating from the bottom of the specimen is effectively the image. For the case of CTEM images, the incident wave is simply described as a plane wave, whilst the simulation of images for STEM is discussed in more detail in section 5.6 and some examples are provided in Chapter 8.

1.5 REQUIREMENTS OF A SPECIMEN

A specimen suitable for study by TEM should obviously be thin enough for electron transmission, and representative of the material about which we wish to draw conclusions. These simple requirements imply that in most cases we must prepare a thin specimen from a larger sample, and in all cases we must assure ourselves that the processes of preparation, mounting and examination do not change, in any uncontrolled way, the important features of the specimen.

For high resolution studies, whether imaging or analysis, the constraints are more severe. The ideal specimen must be thin enough that:

- it can be treated as a weak phase object (see below);
- electron beam spreading within the specimen is negligible;
- (if a nano-particle) it can be supported on a thin substrate with the total thickness still small enough to satisfy the two constraints listed above;

... but also thick enough that:

- it is self-supporting over a region large enough to find the features of interest
- its two surfaces are far enough apart that the material within remains characteristic of the bulk (unless, of course, it is the surfaces in which you are interested)
- surface contamination does not dominate the signal
- there is sufficient signal from scattering events to give statistically significant (i.e. low noise or high enough signal to noise ratio) images or spectra.

These requirements imply that, for most materials to be imaged or analysed at atomic-column resolution, the appropriate thickness will lie in the range up to 50 nm.

It would also be helpful if specimens prepared from the bulk could be perfectly flat and parallel-sided with no contamination or surface amorphised layer, while nano-particles could be of regular shape (so that their thickness is calculable from their projected shape and size). Every specimen should resist both ionization damage and displacement damage by the primary beam.

Almost no real specimens meet all these criteria, but the list of ideal properties should always be borne in mind. Three of the requirements merit some further more quantitative consideration.

A *weak phase object* is a specimen so thin that the electron beam passing thorough it only suffers a modest phase shift, while its amplitude remains effectively unchanged. The approximation that a specimen is indeed a weak phase object is important for much of the treatment in Chapter 5. It should be obvious that such a specimen must be substantially thinner than the mean free path for all the inelastic scattering processes. In practice many biological specimens will meet this criterion at thicknesses below 50 nm, but specimens containing substantially heavier atoms would need to be much thinner.

Beam spreading in a specimen of thickness t is difficult to calculate if multiple scattering is involved, but for specimens which are thin enough to act as weak phase objects and in which single scattering is a good assumption, the beam spreading, b, can be estimated using an equation such as (Goldstein *et al.*, 2003)

$$b = 7.2 \times 10^5 \cdot (Z/E_o) \cdot (\rho/A)^{0.5} \cdot t^{1.5} \qquad (1.3)$$

where E_0 is the electron beam energy in keV, ρ the density in g/cm^3 and Z and A are the atomic number and atomic weight respectively; note that b and t are in cm. You will find that beam broadening is small in a weak phase object but it is worth remembering that for atom column resolution it only requires a broadening of 0.2 nm to take a significant fraction of the beam intensity to the next atom columns. Bear in mind that most specimens are actually thicker than the weak phase object regime, and that beam spreading would be greater than implied by the Goldstein approach. However Goldstein accounts for electrons scattered in all directions, each of which could excite an X-ray. Not all of these electrons could enter an EELS spectrometer, so for EELS the equation might overestimate beam broadening. Strengthening what was said at the beginning of the paragraph, beam spreading is difficult to estimate, never mind calculate!

For a given incident beam energy, beam damage of the specimen is generally a function of the electron fluence (i.e the total number of electrons incident per unit area of specimen) and hence energy deposited within the specimen volume (known as dose); in some cases the fluence rate (usually quoted in current per unit area) can be important. There is some confusion in the literature about these terms; you are advised to read and consider the units of these quantities carefully. Beam damage of the specimen can occur by two dominant mechanisms – knock-on damage in which an atom or ion is displaced from its normal site, and ionisation damage (in some contexts called radiolysis) in which electrons are perturbed leading to chemical and then possibly structural changes. Both mechanisms are discussed by Williams and Carter (2009) and also Egerton *et al.* (2004), who give a chart showing the ranges of primary beam energy which are likely to cause displacement damage for specific atomic species.

Both types of damage are very difficult to predict or quantify with accuracy, because they depend on the bonding environment of the atoms in the specimen. However, in most circumstances, the knock-on cross-section increases with primary beam energy, while the ionisation cross-section decreases. There is thus a compromise to be struck for each specimen to find a beam energy which is low enough not to cause significant atomic displacement but is high enough to suppress too much radiolysis. We return to the subject of beam damage in STEM in section 3.5.

There is no right answer to the choice of specimen thickness or beam energy. Many microscopists only have easy access to a limited range of beam energies and struggle to prepare a good thin specimen, so in practice microscopy involves looking around the specimen for a 'good thin region' and studying it with 100 keV or 200 keV electrons. In the future more consideration may be given to the choice of optimum beam energy, but this requires that the alignment of the microscope and any ancillary equipment such as an EELS spectrometer at any keV be a quick and simple procedure – which it usually is not.

1.6 STEM VERSUS CTEM

There are some advantages to the use of STEM imaging, rather than CTEM. It should be obvious that the STEM configuration is ideal for performing analyses point by point at high sensitivity using multiple signals, and indeed this fact forms the basis for this whole book.

Additionally it should also be apparent that, assuming the signal collection efficiency in STEM is optimised or multiple STEM signals are simultaneously acquired, the total electron fluence or dose which must be delivered to a specimen pixel to generate a specified signal-to-noise ratio (whether for an image or an analytical spectrum) is the same whether delivered by CTEM or STEM. However the dose in STEM is delivered over a short period, which is followed by a longer period of relaxation, whereas in CTEM the instantaneous dose rate is much lower but the dose is continuous. These do not necessarily lead to the same damage, particularly if specimen heating is involved; here the perceived wisdom is that local heating is less for a focused STEM probe than for broad beam CTEM owing to the increased diffusion of thermal energy into surrounding, un-illuminated, cold areas. There is also the possibility, using digital STEM, of positioning the beam consecutively on pixels which are not adjacent – i.e. sampling periodically rather than flooding the whole area. This too can reduce damage at each pixel.

1.7 TWO DIMENSIONAL AND THREE DIMENSIONAL INFORMATION

TEM is a transmission technique and by its very nature produces a two dimensional projection of the interaction of the electron beam with the specimen, whether that be a projected image, a diffraction pattern down a particular crystallographic direction or an analytical signal from a projected through-thickness volume of the specimen. However, in microscopy in general, there is increasing interest in the determination of three dimensional information – known as tomography.

Electron tomography using both CTEM and STEM can be achieved by recording (usually) images for a number of different projections of the specimen and then recombining these images mathematically to form a three dimensional representation of the specimen. This may be achieved in one of three ways: the first, known as *tilt tomography*, is to actually tilt the specimen, usually incrementally but in some cases down certain directions, and record a set of images; the second method assumes that a set of separated objects (e.g. particles dispersed on a TEM support film) are all identical but are arranged over all possible orientations with respect to the direction of the electron beam and is known as *single particle analysis*. The final method known as *confocal electron microscopy* restricts the depth of field/focus in the image to a very thin plane using some form of confocal aperture before the image plane; a set of images are then recorded over a range of defocus and these are combined to give the three dimensional specimen. Chapter 8 gives some recent examples of such STEM tomography and indicates where aberration correction has made a substantial impact.

In conclusion, this initial chapter has both introduced and highlighted some of the important classifications and background theory as well as some of the key issues and developments for TEM in general. As mentioned earlier, the purpose of this handbook, and hence the following chapters, is to focus primarily on the benefits of aberration correction for the formation of smaller electron probes and hence, it will primarily focus on analytical electron microscopy and therefore STEM. However, aberration correction within the context of CTEM is discussed for comparative purposes in Chapter 9.

REFERENCES

Brydson, R. (2001) *Electron Energy Loss Spectroscopy*, Bios, Oxford.

Egerton, R.F. (1996) *Electron Energy Loss Spectroscopy in the Electron Microscope*, Plenum Press: New York.

Egerton, R.F. Li, P. and Malac, M. (2004) Radiation damage in the TEM and SEM, *Micron 35*, 399–409.

Goldstein, J. Newbury, D.E., Joy, D.C., Lyman, C.E., Echlin, P., Lifshin, E., Sawyer, L. and Michael, J.R. (2003) *Scanning Electron Microscopy and X-ray Microanalysis*, Springer.

Goodhew, P.J., Humphreys, F.J. and Beanland, R. (2001) *Electron Microscopy and Analysis*. Taylor and Francis.

Hawkes, P.W. (ed.) (2008) *Advances in Imaging and Electron Physics*, Vol. 153, Elsevier.

Kirkland, E.J. (1998) *Advanced Computing in Electron Microscopy*, Plenum Press, New York.

Koch, C.T. (2002) Determination Of Core Structure Periodicity And Point Defect Density Along Dislocations, PhD Thesis, Arizona State University.
See also http://www.christophtkoch.com/stem/index.html
Stadelmann, P.A. (1987) EMS – A Software Package for Electron Diffraction Analysis and HREM Image Simulation in Materials Science, *Ultramicroscopy* 21: 131–146.
See also http://cimewww.epfl.ch/people/stadelmann/jemsWebSite/jems.html
Williams, D.B. and Carter, C.B. (2009) *Transmission Electron Microscopy: A Textbook for Materials*, Springer.

.

2

Introduction to Electron Optics

Gordon Tatlock

School of Engineering, University of Liverpool, Liverpool, UK

2.1 REVISION OF MICROSCOPY WITH VISIBLE LIGHT AND ELECTRONS

There are many parallels to be drawn between (visible) light optics and electron optics. The major difference, of course, is the wavelength of the illumination used: 450–600 nm for visible light but only 3.7×10^{-3} nm for electrons accelerated through 100 kV. This difference not only controls the ultimate resolution of the microscope but also its size and shape. For example, the scattering angles are usually much smaller in electron optics and rays travel much closer to the optic axis.

Apart from the special case of a lensless projection image microscope, such as a field ion microscope, in which ions from an atomically sharp tip are accelerated across a gap to a large screen, giving a projected image of the source (Miller *et al.*, 1996), most microscopes employ lenses to produce increased magnification of an object. Provided that the object is outside the focal length of the lens, this will lead to a projected, inverted image, which can be used as the object for the next lens, and so on. A typical compound microscope arrangement is illustrated in

Aberration-Corrected Analytical Transmission Electron Microscopy, First Edition.
Edited by Rik Brydson.
© 2011 John Wiley & Sons, Ltd. Published 2011 by John Wiley & Sons, Ltd.

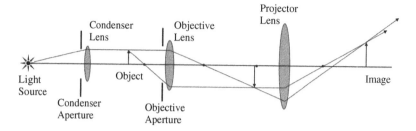

Figure 2.1   Schematic ray diagram of a compound projection microscope used as the basis for a light microscope or a transmission electron microscope.

Figure 2.1 and applies equally to a light microscope or a TEM. Classical lens equations can then be used to link the object and image distances, and hence the magnification, to the focal lengths of the lenses (Hecht, 1998).

As mentioned in Chapter 1, an alternative approach to the use of a wide beam of illumination to image the object (as in TEM) is the use of a fine probe in scanning mode. The object is illuminated point by point in a serial raster and the image is built up pixel by pixel. This is the case for STEM and SEM but also for probe forming techniques such as AFM and STM. However, the magnification, as opposed to the resolution, is now governed only by the electronic signal processing (effectively the area scanned on the specimen). In all cases where lenses are used, the resolution limit is determined not only by the wavelength of the illumination but also by diffraction effects, which limit the number of off-axis beams which can still contribute to the image without distortion, and the aberrations induced by the lens systems.

2.2 FRESNEL AND FRAUNHOFER DIFFRACTION

For most optical systems with apertures of a finite size, the phase of the waves across each aperture is not a linear function of position. When viewed in the near field close to the aperture, these waves need to be treated as spherical waves, not plane waves, with all the complexity that this can generate. In some cases, a sample in the TEM also behaves like an aperture and fringes (Fresnel fringes) appear at sharp edges on the image of the sample and degrade the resolution. The sum over all the different spherical waves – the Fresnel integral – can be treated in a graphical way using an amplitude-phase diagram in which the amplitude is represented by a distance in the xy plane and the phase is represented by angle. By plotting how the overall intensity changes as the waves are summed, it is possible, for example, to illustrate how the intensity changes in the vicinity of a sharp edge in the specimen. It is often marked by a small amount of intensity in the shadow of the edge and an oscillating intensity level close to the edge, which dies away with position away from the edge. Hence the Fresnel fringes.

By adding lenses, however, it is possible to change the situation in some regions of the optical system to one involving only plane waves with effective sources and images at infinity – this is the so-called far-field or Fraunhofer condition, which is a special case of Fresnel (or near field) diffraction. For the case of a plane wave incident on a TEM sample, the scattered waves are brought to a point focus on the back focal plane of the objective lens. For this condition the phases across the aperture are a linear function of position and the phase differences between adjacent secondary waves are all constant. Hence the Fraunhofer diffraction pattern is just the Fourier Transform of the electron wave distribution across the aperture or sample, which we observe in the CTEM as a diffraction pattern.

2.3 IMAGE RESOLUTION

Diffraction effects may also be observed when the image of a point source is formed by an aberration-free convergent lens. Only a part of the wavefront is collected, so that the image is not perfect, and a series of concentric rings is observed (Figure 2.2). Solutions of the integral over the lens and/or aperture involve a mathematical function known as a Bessel function and the intensity oscillates away from the centre of the pattern. The central bright disc is referred to as the Airy disc and the radius of this disc is given by:

$$r_d = \frac{0.61\lambda}{n_r \sin \alpha} \qquad (2.1)$$

Where λ is wavelength, n_r the refractive index of the intervening medium and α is the collection semiangle of the lens/aperture and $n_r \sin \alpha$ is known as the numerical aperture of the lens. The form of this function is also shown in Figure 2.3. Adjacent points in the image would each have a disc associated with them, so one measure of resolution would be how close the points can be brought together, while still being viewed as separate sources. One quite conservative view would be that the two functions may still be resolved when the maximum of one distribution

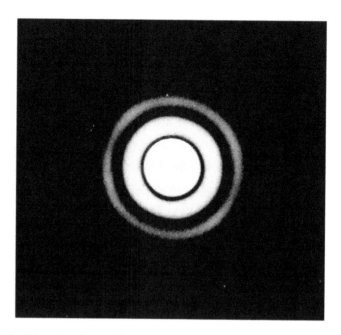

Figure 2.2 Airy ring, formed by white light travelling through a 1 mm diameter aperture. (Reprinted from Eugene Hecht, *Optics*, Copyright 1998, Addison-Wesley-Longman.)

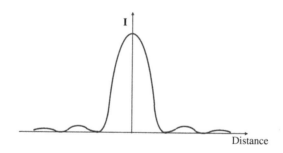

Figure 2.3 Intensity distribution across an Airy ring.

falls on the minimum of the next, i.e. when the separation is the same as the radius of the Airy disc. This is known as the Rayleigh criterion. Clearly it may be possible to separate two distributions which are at a slightly closer distance, but the Rayleigh criterion does have the distinct advantage of simplicity of use.

For an electron lens, n_r can be assumed to be unity (since we are in a vacuum) and $\sin\alpha \approx \alpha$ (when α, the semi-angle collected by the lens, is less than a few degrees). Hence:

$$r_d \approx \frac{0.61\lambda}{\alpha} \tag{2.2}$$

2.4 ELECTRON LENSES

Both electrostatic and electromagnetic lenses have been used in electron microscopes, although electrostatic lenses can have larger aberration coefficients which are prone to change with time, as insulating films build up or microdischarges occur on the electrodes. However electrostatic lenses do have the advantage of being rotation-free (i.e. they do not induce rotation of the image relative to the object) and are usually easy to construct. They are rarely used in the imaging systems of present day electron microscopes, although most illuminating systems still contain a series of electrostatic lenses which extract or accelerate the electron beam from the gun filament or tip.

Rotationally symmetric electromagnetic lenses consist of a wound coil through which a current is passed. Electrons passing through the coil undergo a focusing effect which is dependent on the coil current and which can be increased by increasing the concentration of the resultant magnetic field with a soft iron pole piece. The small air gap in this pole piece concentrates the field and shortens the focal length of the lens (Figure 2.4). One limit to the strength of the lens is the saturation of the magnetisation, which can be overcome to some extent by using a lens in

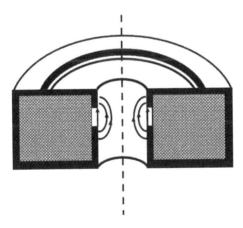

Figure 2.4 A schematic (cut-away) diagram of a rotationally symmetric electron lens.

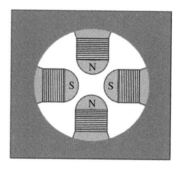

Figure 2.5   Schematic diagram of a quadrupole lens configuration.

which the iron alloy pole pieces are replaced by components containing rare-earth materials with a higher saturation magnetisation. Alternatively, hollow cylinders of superconducting material can be included within a conventional lens to confine the magnetic flux into a particular configuration by screening.

Other types of lenses sometimes found in electron microscopes include the mini-lens (Mulvey, 1974) which has a water cooled coil or snorkel lens with only a single pole piece. Small bores allow a reduction in focal length and spherical aberration coefficients. We can also get multipole lenses such as quadrupole lenses constructed from four pole pieces of opposite polarity (Figure 2.5). In this case the magnetic field is perpendicular to the electron beam and a strong Lorentz force is exerted. They can be used to correct or compensate for aberrations in the main rotationally symmetric electromagnetic lenses (see Chapter 4).

2.4.1 Electron Trajectories

The force $\mathbf{F}$ on an electron with charge $-e$ travelling with velocity $\mathbf{v}$ in a magnetic field $\mathbf{B}$ is given by the vector equation:

$$\mathbf{F} = -e(\mathbf{v} \times \mathbf{B}) \qquad (2.3)$$

in which the magnitude of $\mathbf{F}$, $|\mathbf{F}| = |\mathbf{B}||e||\mathbf{v}| \sin\theta$. Where θ is the angle between B and $\mathbf{v}$. Hence if a component of $\mathbf{v}$ is parallel to $\mathbf{B}$, $|\mathbf{F}| = 0$, but any perpendicular component will result in a force on the electron, which will describe a circular motion about $\mathbf{B}$. The resulting path of the electron will describe a helix. If the cylindrical lens has an homogeneous field then the electrons which pass exactly through the centre will experience no force but those some distance off the main axis will spiral towards the centre and then out again, leading to a focusing action of the lens. The focal length of the lens is governed by the magnitude of $|\mathbf{B}|^2$ and hence the current applied to the lens coil. Combinations of lenses can give a variable range of image magnifications while correcting for any image rotation, which would occur using just a single lens.

2.4.2 Aberrations

In an ideal optical system every point on the object will be reproduced perfectly in the image. In practice, this is unlikely to be true for all but paraxial rays which are very close to the optic axis of the system (within 100 µm and subtending an angle of less than 10 mrad). For lenses with paraxial rays, Gaussian or first order optics applies (see section 4.2). If the angle a ray makes with the optic axis is θ then for very small angles $\sin\theta \sim \theta$ and $\cos\theta \sim 1$, a first order approximation. However, if $\sin\theta$ is written as a series expansion, then:

$$\sin\theta = \theta - \frac{\theta^3}{3!} + \frac{\theta^5}{5!} \cdots\cdots \tag{2.4}$$

Hence the next level of approximation would be third order. At this level many of the distorting effect of the lenses start to become more obvious in the form of aberrations, of which there are five basic types. The monochromatic aberrations include spherical aberration, astigmatism, coma, field curvature and distortion. In addition there are also chromatic aberrations which arise when the electron beam is not monochromatic. This may be due to the inherent energy spread in the electrons emitted from a source, fluctuations in the gun accelerating voltage or different energy losses due to interactions with the sample (see Chapter 1), for example. Waves of different wavelengths (analogous to colours in visible light) are then focused at different points on the optic axis, due to the chromatic aberrations in the lenses (Figure 2.6).

Of the five monochromatic aberrations, the most important are spherical aberration, astigmatism and coma. These aberrations are discussed in detail in Chapters 4 and 9. Diagrams illustrating the effects of spherical aberration and astigmatism are shown in Figure 2.7. The key point about spherical aberration is that off-axis rays are not brought to a focus at the same point as paraxial rays. With longitudinal spherical aberration, the different cross-overs still lie on the optic axis but transverse spherical aberration can also occur. In the case of astigmatism, the cross-over of the off-axis rays is displaced along the optic axis but now

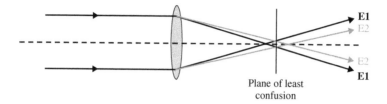

Figure 2.6 Chromatic aberrations in the lenses result in waves of different wavelengths and energies (E1 < E2) being brought to a focus at different points along the optic axis. The disc of least confusion marks the plane where the effects of the aberration are minimised.

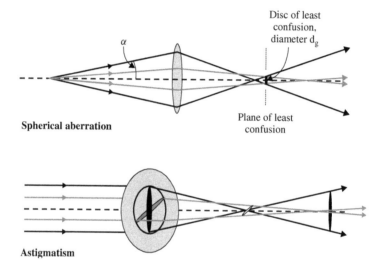

Figure 2.7 Two of the five different sources of monochromatic aberration: positive spherical aberration and astigmatism. α is the collection semi-angle of the lens and d_g is the diameter of the disc of least confusion.

the degree of displacement varies with the azimuthal angle of the beam under consideration.

Coma or comatic aberrations are related primarily to rays which are emitted from an object point slightly off the optic axis. All the rays which travel through the centre of the lens will still be brought to a point focus off the optic axis but rays which travel through the peripheral field of the lens will be focussed at different points (see figure 2.8). Any anisotropy in the lenses also adds additional contributions to coma, astigmatism and distortion.

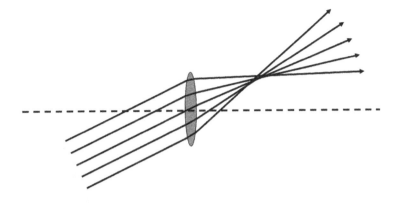

Figure 2.8 Coma or comatic aberration in which off-axis peripheral rays are brought to a different focus from those that travel through the centre of the lens.

The degree of spherical aberration of a lens can be related to the diameter of the disc of least confusion d_g and the collection semi-angle of the lens, α, defined in Figure 2.7, by:

$$d_g = \frac{1}{2}C_s\alpha^3 \tag{2.5}$$

where C_s is the spherical aberration coefficient; note this varies as the cube of the angle of the ray to the optic axis, θ (Goodhew *et al.*, 2001). Similarly the chromatic aberration coefficient C_c can be related to the disc diameter d_c and energy spread ΔE by:

$$d_c = C_c\alpha\,\Delta E/E_o \tag{2.6}$$

where E_o is the energy of the primary beam (Goodhew *et al.*, 2001). Hence for both of these effects the resolution will degrade as α becomes larger. However equation (2.2) using the Rayleigh criterion, suggests that the resolution improves as α increases. Hence if there is a combination of both diffraction and spherical aberration effects limiting the resolution, there must be an optimum figure for resolution as α changes. For example, if the spherical aberration and diffraction limit alone are considered and the net resolution (r_{total} – the radius of the total disc of confusion) were just the sum of r_d and r_g, with r_g proportional to $C_s\,\alpha^3$, then minimizing r with respect to α gives:

$$r_{total} = 1.21\lambda^{3/4}C_s^{1/4} \tag{2.7}$$

which is often quoted as a simple guide to microscope resolution. A more complete treatment of all these features is given in Chapters 4 and 6.

2.5 ELECTRON SOURCES

A range of electron sources or electron guns is available for use in electron microscopes. The choice depends largely on the beam current and energy spread that are required, together with the vacuum which can be maintained in the gun chamber. One quantity that is often quoted is the gun brightness (B), which is defined as the current density (j) per unit solid angle $(d\Omega)$ i.e.

$$B = \frac{j}{d\Omega} = \frac{j}{\pi\alpha^2} \tag{2.8}$$

where j is the current density and α is the semi-angle of the cone of electrons emerging from the surface of the gun.

In all cases, electrons have to be extracted or emitted from a tip, which may be heated to encourage emission. Hence the choice of material for the tip is usually restricted to high melting point materials, such as tungsten, and/or low work function materials such as LaB_6, the latter owing to that fact that if conduction band electrons have to escape from a solid surface into a vacuum, they must overcome a potential barrier or work function at the surface.

A heated hair-pin filament made of a bent piece of tungsten wire is the simplest and cheapest form of thermionic emitter which can give a high beam current under moderate vacuum conditions. A typical arrangement is shown schematically in Figure 2.9. The filamentary

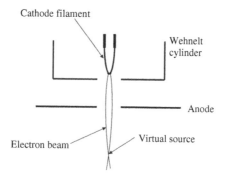

Figure 2.9   Schematic diagram of a thermionic electron gun.

cathode is surrounded by a Wehnelt cylinder which is maintained at a potential of a few hundred volts with respect to the cathode. This is used to focus the electrons, which are then accelerated across a high potential difference to the anode. The total emission current can be as high as $10-100\,\mu A$, but the energy spread of a $100\,keV$ beam may be several electron volts. Hence thermionic emitters are not suitable for use with high energy resolution EEL spectrometry for example, where we require the energy resolution to be ca. $0.3\,eV$.

If a large total beam current is a high priority, for EDX analysis, for example, one alternative is the use of a LaB_6 pointed cathode, which is heated indirectly. This has the advantage of a factor of times 10 increase in beam brightness and beam current, with a slightly reduced energy spread. However, in order to achieve an energy spread of below $0.5\,eV$ with a high beam coherence, a cold field emission tip is usually required. In this type of gun, a single crystal tungsten tip, usually in a <310> or <111> orientation, is used with a double anode configuration. The first anode, sitting at a potential of about $4\,kV$ with respect to the tip, is used to extract the electrons from a very small region of the cold tip. Hence the overall brightness is very high (10^9 Amp cm^{-2}/per unit solid angle (Steradian, sr)), although the total current emitted is relatively low ($1-10\,\mu A$). Care is also needed with the design of the first anode/tip configuration to ensure that extra aberrations are not introduced and that the energy spread of the beam is not degraded by the Boersch effect (bunching of electrons) (Boersch, 1954).

Another problem with this type of arrangement (shown in Figure 2.10) is that the emission decreases with time due to changes in the tip and the arrival of impurity atoms on the surface. Hence the tip needs to be flash heated periodically to remove any impurities and restore the emission level. Notwithstanding, in order to achieve the best field emission tip stability, the best vacuum possible is required in the gun chamber. In many cases a simple gun lens is used to couple the gun to the subsequent condenser lens system (see later and also Chapter 6); this provides an

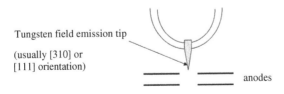

Tungsten field emission tip
(usually [310] or
[111] orientation)

anodes

Figure 2.10 Schematic representation of a tungsten single crystal field emission electron gun.

Table 2.1   Typical figures for brightness, current and energy spread for various types of electron guns

	Tungsten hair pin	LaB$_6$	Tungsten cold field emission
Brightness (Acm^{-2}sr^{-1})	10^4	10^5	10^9
Total current emitted (μA)	10–100	100	1–10
Energy width (eV)	1.0–2.0	0.5–2.0	0.1–0.5

extra benefit in that positive ions, which may otherwise reach the tip from the column, may be deflected by the gun lens and have much less effect on overall performance of the gun.

An attractive alternative to the use of a cold field emitter is the so called warm field emitter or Schottky emitter, in which the tip is warmed so that a smaller extraction voltage is needed and also the tip emission is more stable with time. The field produced is enough to lower the work function and increase the total emission when compared to a cold field emitter. This may be at the expense of a slight increase in energy spread and loss of beam coherence, in comparison with the cold field emitter. The figures for all the different types of tips are summarised in Table 2.1.

2.6 PROBE FORMING OPTICS AND APERTURES

Most probe forming systems have a multiple lens condenser system so that both the probe size and probe convergence angle can be varied independently. In some cases a gun lens is also used.

Using the expression for brightness (B) in equation (2.8), conserving gun brightness along the optic axis and neglecting the fact that the probe is likely to have a Gaussian distribution, then the total current (I) into a spot of diameter d$_s$ (source diameter) would be:

$$I = \frac{\pi d_s^2}{4} j \quad \text{where } j = \pi \alpha^2 \cdot B \qquad (2.9)$$

Hence

$$d_s = \frac{2}{\pi} \left(\frac{I}{B} \right)^{\frac{1}{2}} \cdot \frac{1}{\alpha} \qquad (2.10)$$

Note that for a given brightness (B), as the probe/spot size (d$_s$) is decreased then the probe convergence angle (α) increases and therefore the probe current (I) will remain the same, if the aberrations are corrected. This will be revisited in Chapter 6 after a more detailed discussion of aberration correction in Chapter 4.

2.7 SEM, TEM AND STEM

In section 1.1. we introduced the various operational strategies for electron microscopes and here we reiterate them in a little more detail.

In a conventional scanning electron microscope (SEM) an electron probe is focussed by an objective lens onto the surface of a sample and this probe is scanned in a raster across the surface. The secondary electron signal from the entrance surface is collected by a detector and displayed in a similar raster on a cathode ray tube (CRT) or frame store. Hence scanning the probe over a smaller region on the sample, while maintaining the same size raster on the CRT, will lead to magnification (Figure 2.11). However there is a limit to the minimum size of the probe, and hence the resolution, achievable for a certain value of the probe current, and hence contrast level which is required in order to image a particular feature, as explained in the last section and in Chapter 1.

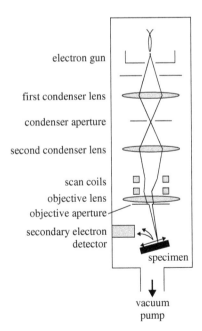

Figure 2.11   Schematic layout of a conventional scanning electron microscope.

The CTEM, on the other hand, works on a totally different principle. The beam is simultaneously illuminated over an area of the sample and the objective lens and subsequent projector lenses are used to form an image of the sample on the final viewing screen or detector array, in much the same way as an optical microscope (Figure 2.12). This is termed the CTEM operating in *image mode*. However, if a parallel beam of electrons is incident on the sample, then beams of electrons scattered in different directions by the sample will be brought to point focii in the back focal plane of the objective lens, giving rise to a diffraction pattern which can also be projected onto the viewing screen or detector array by changing the excitation of the projector lenses after the objective lens and operating the CTEM in *diffraction mode*. If we initially consider only elastic scattering then, in a single crystal sample, each diffraction spot will comprise electrons which have been scattered through a given angle by a set of planes of atoms while a fine grained polycrystalline sample will give rise to a series of concentric rings. Hence this is a very convenient way of obtaining crystallographic information about a sample in addition to images.

If the condenser system, or the pre-field of the objective lens is used to converge the primary beam onto a small region of the sample, then a series of discs are formed in the back focal plane of the objective lens instead of the spots obtained with more parallel illumination. This is the so-called convergent beam pattern, which may be used to obtain extra crystallographic information and symmetry information about a sample (Williams and Carter, 2009). However, if this convergent beam is scanned across the sample in a similar way to SEM imaging, and the transmitted beams are incident on a detector we have the basis of a scanning transmission electron microscope or TEM/STEM (Figure 2.12).

Of course if no diffraction information is required, then all the post specimen lenses in the CTEM could be dispensed with and the objective lens itself used to converge the beam on to the sample, forming a dedicated STEM as shown in Figure 2.13. The post specimen lenses are replaced by a series of detectors or spectrometers, and a combination of the signals from these detectors can then be used to optimize the image contrast. For example, since the high angle scattering is very sensitive to atomic number (see chapter 1), a high angle annular dark field (HAADF) detector is very useful for the generation of direct atomic contrast images (see Chapter 3). Figure 2.14 shows a photograph of

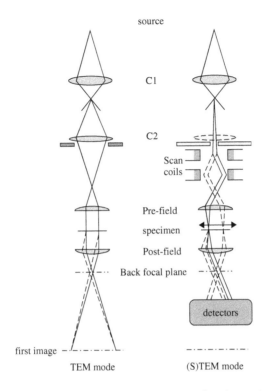

source

C1

C2

Scan
coils

Pre-field

specimen

Post-field

Back focal plane

detectors

first image

TEM mode   (S)TEM mode

Figure 2.12   Comparison of TEM and STEM mode (after Reimer, 1989).

a cutaway STEM currently in the Cavendish Museum at Cambridge together with a corresponding schematic diagram.

In fact, in principle, it is possible to link the imaging operation of the TEM and the STEM, which is known as the principle of reciprocity. This is discussed in more detail in Chapter 3, section 3.3 and also Chapters 5 and 6.

Although, in principle, post specimen lenses may not be required after the sample in a STEM, it is often convenient to have some form of transfer lens after the sample so that, for example, the acceptance angle for an EEL Spectrometer can be optimised, as discussed in Chapter 6. Thus the main purpose of any post specimen lens is as a coupling lens to other detectors or spectrometers (see Chapter 3). A Ronchigram camera may also be inserted, post-specimen, and used to optimise the microscope alignment including any aberration correctors (see Chapters 4, 5 and 6).

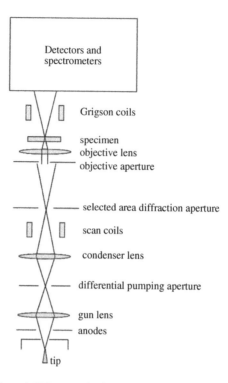

Figure 2.13 Dedicated STEM with the electron gun at the base of the column. There are no post specimen lenses other than transfer lenses to facilitate the use of spectrometers, etc.

In terms of the correction of spherical aberration in both STEM and CTEM, correctors in both types of instruments are based on the principle of introducing negative aberrations to cancel (or adjust) the positive aberrations introduced by the objective lens. More details are provided in subsequent chapters (section 3.6, and Chapters 4, 5, 6 and 9). In a dedicated STEM, since the main function of the objective lens is to define the probe reaching the sample, a multi-multipole lens probe aberration corrector is inserted between the condenser system and the (probe forming) objective lens (see Figure 2.14). In the CTEM however, a similar corrector is usually located after the objective lens giving an image corrector, and this will be discussed in more detail in Chapter 9.

In conclusion, this chapter has highlighted some of the important concepts in electron optics and electron microscopy that we will return to in more detail in some of the subsequent chapters. Initially however, we consider the general development of the STEM in more detail in the following chapter.

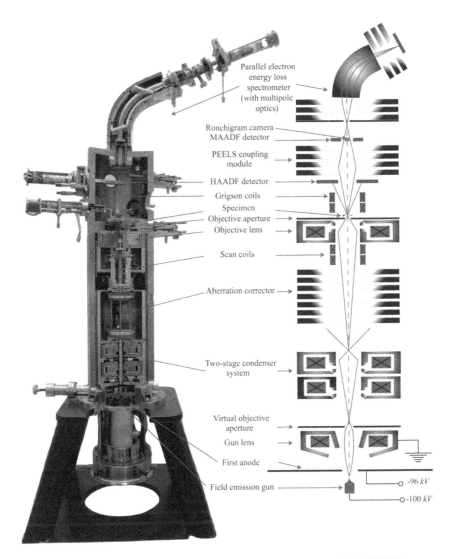

Parallel electron
energy loss
spectrometer
(with multipole
optics)

Ronchigram camera
MAADF detector

PEELS coupling
module

HAADF detector

Grigson coils
Specimen
Objective aperture
Objective lens

Scan coils

Aberration corrector

Two-stage condenser
system

Virtual objective
aperture

Gun lens

First anode

Field emission gun

-96 kV

-100 kV

Figure 2.14 Photograph of a sectioned Vacuum Generators HB5 STEM incorporating an early prototype Nion MarkI quadrupole-octupole C_s corrector in the Cavendish Museum plus corresponding schematic diagram. The original microscope, the second one manufactured by VG, was sectioned in the Cavendish Workshop by David Clarke (courtesy: David Clarke, J.J. Rickard and Quentin Ramasse). (A colour version of this figure appears in the plate section)

REFERENCES

Boersch, H. (1954) Experimentelle Bestimmung der Energieverteilung in Thermisch Ausgelösten Elektronenstrahlen, *Z. Phys.* 139, 115–146.

Goodhew, P.G., Humphreys, F.J. and Beanland, R. (2001) *Electron Microscopy and Analysis*, Taylor and Francis, New York.

Hecht, E. (1998) *Optics* (3rd edn.) Addison-Wesly-Longman, Reading MA.

Miller, M.K., Cerezo A., Hetherington M.G. and Smith G.D.W. (1996) *Atom Probe Field–ion Microscopy*, Oxford University Press, Oxford.

Mulvey, T. (1974) in *Electron Microscopy 1974*, Vol. 1, J.V. Sanders and D.J. Goodchild (eds), Australian Acad. Sci, Canberra.

Reimer, L. (1989), *Transmission Electron Microscopy* (2nd edn), Springer-Verlag, Berlin.

Williams, D.B. and Carter, C.B. (2009) *Transmission Electron Microscopy* (2nd edn), Springer, New York.

3

Development of STEM

L.M. Brown

Robinson College, Cambridge and Cavendish Laboratory, Cambridge, UK

3.1 INTRODUCTION: STRUCTURAL AND ANALYTICAL INFORMATION IN ELECTRON MICROSCOPY

The purpose of this chapter is to present in concise and readable form an outline of the development of scanning transmission electron microscopy.

It is remarkable to think that until 1933 (Ruska, 1986) no one had seen a detail of an object smaller than about half a micron – 500 nm. Nowadays, the entire field of nanotechnology occupies the domain from atomic size to about 100 nm. This development, confidently expected to become a major economic activity impacting on all aspects of life, depends utterly on microscopy. Electron microscopy, with its unique capability to provide both chemical and structural information throughout a nanometer-sized volume, plays a central role in the implementation of nanotechnology.

If one imagines an atom-by-atom description of a structure, one sees that each atom is specified by its co-ordinates and its chemical type and state, for example its ionic charge. This information is in principle conveyed by an electron energy-loss spectrum obtained by a

Aberration-Corrected Analytical Transmission Electron Microscopy, First Edition.
Edited by Rik Brydson.
© 2011 John Wiley & Sons, Ltd. Published 2011 by John Wiley & Sons, Ltd.

probe focused on the atom (Chapter 7). In such a spectrum, the joint
density of electronic states in energy – a convolution of both empty
and full states – is determined from the low-loss part, and the density
of empty states is determined from core losses. Deconvolution of the
low-loss spectrum to separate occupied from empty states can then be
contemplated, and full information obtained. The information would
be sufficient to calculate binding energies directly from the distribution
of bonding electrons. Such an ambitious programme has not yet been
carried out in practice, for a variety of reasons. But it is instructive
to think of the amount of information required. In a particle of size
30 nm, there are typically a million atoms. Each electron energy loss
spectrum, covering a range of energy losses from zero to 2 kV at a
resolution of 0.2 V requires at least ten thousand numbers of variable
magnitude up to one million, if the statistical noise is to be kept below
0.1%. The total information required is a minimum of 10^{16} bits, say
1 Tbyte. Even if the problems of mechanical stability and radiation
damage in the microscope could be overcome, the problem of handling
the data is formidable, requiring processing speeds and random access
memories perhaps a hundred times greater than those now routinely
available. Nevertheless, one can imagine that in the future this might
become possible.

But even if it becomes possible, science does not work in this way.
Blind processing of data without reference to the details of the problem
in hand does not convey understanding. So techniques have evolved
to make progress with important problems such as the structure of
catalyst particles, or of grain boundaries in the solid state, without
the acquisition of a surfeit of data. Even so, it is difficult to extract
both spectral information and structural information simultaneously.
Early attempts to do this used a mechanical slit to select a line in the
image of the sample from which energy-loss spectra could be obtained
(Metherell and Whelan, 1965). The main problem with this method was
mechanical construction and control of the slit. It was realised very early
on that the simplest way forward was to acquire images and spectra
simultaneously point-by-point, but the electron guns available then were
not bright enough to do this in an acceptable time. In a conventional gun
with a thermionic filament (Figure 2.9), the source of the electron beam
is almost of millimetre dimensions, and must be hugely demagnified
if a beam of nanometre dimensions is to be obtained. The solution
was seen to be the development of field emission guns (Figure 2.10),
where the source of the current is of atomic dimensions. But such guns,
based on electron tunneling, require upgrading the vacuum system of

microscopes to ultra-high-vacuum (UHV) standards, technically very difficult and costly.

3.2 THE CREWE REVOLUTION: HOW STEM SOLVES THE INFORMATION PROBLEM

The first implementation of point-by-point acquisition of spectra and the imaging of individual atoms on a substrate was carried out by Albert Victor Crewe (Crewe, 1971) in 1970. His instrument was the first with an electron probe capable of dwelling on each pixel in the image and thereby producing simultaneously image information (elastic scattering) and chemical information (inelastic scattering), both limited only by the electron optics and the stability of the instrument over the acquisition time. Novel features of the instrument included not only ultra-high vacuum, and a field emission electron gun, but also a unique extraction lens in the gun which corrected aberrations, and the invention of the 'annular dark field detector', or 'ADF'.

Crewe's ADF produced images with high contrast. The scattering into the ADF comes mostly from Rutherford scattering by the atomic nuclei sampled by the beam (see section 1.2.1). Thus the intensity in the image increases both with the atomic number of the nuclei and the actual number of them sampled by the beam. An improvement was to divide the intensity at each image point by the intensity in the inelastic signal, which increases with the thickness of the sample. The resulting signal displays intensity approximately proportional to the atomic number, independent of foil thickness, and was called by Crewe 'Z-contrast'. Using Z-contrast Crewe imaged single atoms of heavy metals on amorphous substrates.

Electron microscope groups at that time were unaccustomed to ultra-high vacuum. Crewe, with his background in accelerator physics, could tackle both the vacuum problems and novel electrostatic methods of focusing the beam. He is very much a 'particle' man and not a 'wave' man, so he recognised that electrons scattered to large angles are deflected mainly by the atomic nucleus with its concentrated charge, whereas those deflected to small angles are mostly inelastically scattered by the more widely distributed atomic electrons. He therefore designed an annular detector which collected very efficiently the electrons scattered to large angles, while at the same time it allowed the beam to pass through a central hole. Simultaneously in another axial detector he collected those deflected to small angles after they passed through the hole in the annular detector and then through a magnetic prism spectrometer which separated those which had lost energy from those which were only elastically scattered. The high-angle signal he assumed proportional to the square of the atomic number Z. However, the inelastic signal is more nearly proportional to Z itself. For thin samples, both signal strengths increase proportional to the sample thickness, so the ratio between them he called 'Z-contrast'. The ratio enabled intensity variations due to changes in specimen thickness to be factored out, leaving only variations due to the atomic number of the atoms in the pixel. This is what enabled him to produce the first pictures of isolated heavy atoms on an amorphous substrate.

At about the same time, Isaacson and Johnson (1975), members of Crewe's group at the University of Chicago, closely followed by Leapman and Cosslett (1976) at Cambridge, implemented the first effective quantitative microanalysis using EELS. Although their methods have now been largely superseded by built-in software, they recognised that the spectra are basically absorption spectra and utilised power-law extrapolation to find the background which must be subtracted to obtain the counts under an edge, a procedure still routinely used to perform microanalysis by EELS (see section 7.5.1).

The basic configuration of Crewe's instrument, so revolutionary at the time, persists in all modern dedicated STEM (see Figure 2.14).

3.3 ELECTRON OPTICAL SIMPLICITY OF STEM

Scanning transmission electron microscopy can be regarded as a kind of 'radar'. A beam scans the region of interest. Its position is slaved point by point to a 'raster' scan (the word 'raster' probably derives from the Latin word for mattock or rake: an implement for scratching the earth in lines) which displays the intensity of a selected signal, for example a reflection from the beam. Such an imaging mechanism was regarded as 'incoherent', meaning that phase information is lost. You can think of the electrons as particles, some of which are bounced into the detector. It came therefore as a surprise that electron optical interference effects could be seen in STEM. These come about because of the reciprocity theorem, which says that if the scattering of a beam is elastic, identical images are obtained if the source and detector are interchanged (Cowley, 1969).

Figure 3.1 shows a simplified diagram (Brown, 1981). In Figure 3.1(a), an electron source illuminates a specimen, and an objective lens forms a

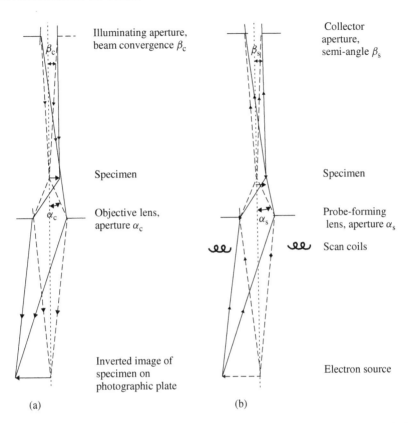

Illuminating aperture, beam convergence β_c

Collector aperture, semi-angle β_s

Specimen

Specimen

Objective lens, aperture α_c

Probe-forming lens, aperture α_s

Scan coils

Inverted image of specimen on photographic plate

Electron source

(a)

(b)

Figure 3.1 (a) shows a single-lens conventional microscope (CTEM) and (b) a single-lens scanning transmission microscope (STEM). In (a) the source is an extended source illuminating the whole of the object; rays from two of the point sources making up the extended source are shown. In (b), the true source is very nearly a point source whose effective position is changed by the action of the scanning coils; two positions are shown. In this particular diagram, the angle β in CTEM refers to the beam convergence, determined by the condenser lenses; in STEM it refers to the collector aperture. The angle α in CTEM refers to the objective semi-aperture angle, and in STEM to the aperture of the probe-forming lens. Reprinted from L. M. Brown, Scanning transmission electron microscopy: microanalysis for the microelectronic age, *J. Phys. F: Metal Physics*, Vol. 11, p1–26, Copyright 1981 with permission of IOP.

conventional image which is recorded on a CCD camera. In Figure 3.1(b), a lens is used to form a probe which is made to scan the specimen. Because in the scanning arrangement of Figure 3.1(b) the specimen is illuminated point-by-point, a variety of detectors can be used to obtain information. In the conventional set-up of Figure 3.1(a), the specimen is illuminated over a wide area, and point-by-point information is obtainable only from individual pixels of the CCD camera.

It is clear in Figure 3.1 that the ray paths (full lines) of Figure 3.1(a) are identical to those of Figure 3.1(b), except that the source and detector positions are interchanged. It follows that the amplitude at any point in the collector due to unit amplitude at the source S is the same as the amplitude at S due to the unit amplitude imagined to emanate from a point in the detector. If it is assumed that all points in the source in Figure 3.1(a) emit incoherently, and that all points in the detector detect incoherently, then the image formed by STEM is identical to the image formed by CTEM, provided the relevant apertures are the same.

Thus in STEM the resolution in the image is controlled by the size of the probe formed by the lens, limited by lens aberrations and the aperture. In CTEM, the resolution is controlled by the quality of the objective lens, again limited by its aberrations. The 'probe-forming' lens and the 'objective' lens play identical roles in determining the electron-optical resolution limit in both instruments.

In addition to the concept of the resolution element, it is necessary to think of the picture element which can be defined either by reference to the size of detail of interest in the specimen, or by reference to the recording system of the microscope. This is determined by the separation of pixels, usually in the CCD camera.

One other essential factor is noise in the image, which if the electronics is perfect is determined fundamentally by the number of electrons per picture element, n. The noise to signal ratio is approximately $n^{-1/2}$ and this must be much less than the corresponding contrast variation between picture elements (see section 1.3). The contrast at a picture element, defined as the fractional change in signal between the chosen element and its neighbours, must be about twenty times the fractional standard deviation, if it is to be reliably detected. This means that to see a contrast of 2% requires about a million electrons per picture element. Such a crude calculation however does not take into account the ability of the eye to see recognisable shapes; a faint bright line can be discerned even in a very noisy image. Nevertheless, a very good image is formed from 10^6 electrons per picture element, and a useful one, especially of periodic or highly structured objects, from 10^4 or so.

This is where the brightness of the electron source comes in. Before aberration correction was available, a high-quality system delivered about 1 nA of current into a picture element 1 nm in size; this amounts to about 5000 million electrons per second. Thus it is possible to acquire high quality images, and simultaneously the corresponding spectra, from tens of thousands of pixels in a few seconds. With aberration correction, the system delivers about the same current into a picture element 0.1 nm or 1Å in size. The old thermionic sources were at least ten thousand times less bright, so that the acquisition times were impossibly long, taking several hours.

3.4 THE SIGNAL FREEDOM OF STEM

After electrons leave the specimen in STEM, there need be no further lenses, only detectors. This simplicity of electron optics enables a wide range of signals to be observed, and appropriate ones chosen to suit the problem in hand. We list those most commonly used:

3.4.1 Bright-Field Detector (Phase Contrast, Diffraction Contrast)

By the principle of reciprocity, the angular size of the bright-field (axial) detector in STEM corresponds to the illuminating aperture in CTEM. If this is smaller than a Bragg angle in a crystalline specimen, diffraction contrast can be observed. The highly-developed theories of diffraction contrast can be taken over without modification. Similarly, phase contrast can utilised. A critical test of the microscope is the number of Fresnel fringes which can be observed at an edge or around a hole in the sample. This tests the source size and the detector size, as well as the mechanical stability of the microscope over the acquisition time for the image. Phase contrast is much more efficient in CTEM than in STEM, because of parallel collection of the image. What is extraordinary is that fringes can be seen at all in a scanning system. This is discussed further in Chapters 5 and 6.

3.4.2 ADF, HAADF

Crewe's annular dark field detector is an efficient way to make use of electrons scattered to large angles. It was particularly effective when used to image isolated heavy atoms on an amorphous substrate. But when applied to crystalline samples, the detector collected a mixture of Bragg diffracted beams and incoherently scattered electrons, producing images that were difficult to interpret. Howie recognised that if the inner angle of the ADF were made large enough, Bragg reflections would be suppressed by thermal vibrations. With his students (Treacy *et al.*, 1980) he conceived and demonstrated the 'High-Angle-Dark-Field-Detector' which almost entirely eliminates diffraction contrast and permits the imaging of heavy atoms on (or in) crystalline materials. HAADF has become the most widely-used imaging method in STEM, because of its efficiency and the ease of interpretation of the images it produces. By reciprocity, it is equivalent to hollow-cone illumination in CTEM, but is much more easily implemented in STEM where there is such freedom to place detectors on the electron-exit side of the specimen. Again further discussion of STEM dark field imaging is given in Chapters 5 and 6.

3.4.3 Nanodiffraction

If transmitted electrons are collected from the focused probe, a diffraction pattern is obtained, equivalent to Convergent Beam Electron Diffraction (CBED) in CTEM. In STEM, these patterns can be collected serially by post-specimen scanning coils, called Grigson coils after their inventor (Grigson, 1965) but in modern instruments they can be collected on a CCD camera after passing the entire diffraction pattern through an electron spectrometer and filtering it to remove inelastic scattering. Such patterns can be used to determine sample structure and thickness, exactly as CBED patterns (see section 8.3.4). The high brightness of the field emission source enables patterns to be collected from a fully focused probe, with a probe size of 0.1 nm and a convergence angle of 30 mr or so. If a conventional source is employed, the probe size must be nearly 100 nm, so the particle must be defect free over at least this volume if a sharp pattern is to be obtained. Greatly improved CBED performance is a by-product of field emission.

In STEM, it is possible to collect CBED patterns from each probe position. Each pattern contains structural information, such as lattice periodicities, coded in the diffraction pattern. In CBED, where the discs of diffracted electrons overlap the relative phases of the diffracted beams can be determined. It follows that the information in a series of CBED patterns is complete structural information, provided a suitable algorithm can be devised to extract it. This method of structure determination by diffractive imaging is called 'Ptychography' by Rodenburg (2008) and promises to produce remarkable advances in microscopy because it does not depend upon the resolution of the microscope, but only on rather complex data handling.

3.4.4 EELS

It was recognised early in the development of electron microscopy that electrons passing through a thin film should suffer energy losses in quanta which are characteristic of its chemical makeup. Early attempts (Ruthemann, 1941) during the Second World War demonstrated the feasibility of the idea, but were hampered by poor vacuum and inadequate brightness of the electron gun. Crewe and his team were the first to demonstrate useful high-resolution results. The technique was slow to catch on because many electron microscopists were unfamiliar with the interpretation of what were rather noisy spectra. An important turning point was the publication by Ahn and Krivanek (1983) of the 'EELS Atlas' which contained a complete set of reference spectra; also the availability of commercial spectrometers. Egerton's pioneering book (Egerton 1996) brought the concepts and techniques to all laboratories concerned with electron microscopy.

EELS produces analysis at the highest possible spatial resolution, limited only by the probe size and the 'impact parameter': the closest distance of approach between a passing fast electron and a target atom. The electron causes a pulse of electric field which induces the transition, but only if it is close enough that the pulse contains frequencies high enough. The simple rule for fast electrons of 100 kV energy is that the maximum impact parameter capable of inducing an energy loss of E is $(40/nm)/(E/eV)$. Thus the maximum impact parameter for the carbon K edge at 285 eV is 0.14 nm, whereas for a low plasmon loss of 20 eV it is 2 nm. For losses above about 100 eV EELS becomes an analytical technique capable of genuinely atomic resolution. Recently, EELS from single embedded atoms have been detected (Krivanek *et al.*, 2008).

3.4.5 EDX

Once the inelastic transition has taken place, causing the fast electron to lose energy by producing a hole in the distribution of atomic core electrons, the hole is healed by the transition of an outer electron into it. An X-ray and/or Auger electrons are produced. The characteristic X-ray energy reveals the chemical nature of the atom. The X-ray spectrum is an emission spectrum, a peak on a low background, much easier to analyse than the energy-loss in EELS. Commercial energy-dispersive X-ray (EDX) detectors are used to display the spectra.

Despite their apparent simplicity, EDX spectra suffer from several disadvantages. They yield only a slow count rate, because only a fraction of the electron energy losses produce X-rays, and because the detectors subtend only a small solid angle at the specimen so the collection efficiency is very poor. The spectra contain no detailed information on the density of electronic states seen by the struck electron. The greatest drawback is that the X-ray can be emitted from anywhere in the electron-illuminated area, so that the spatial resolution attainable by the technique is determined by the probe size as it spreads out inside the sample. For a foil of 30 nm thickness, the exit probe size is 1 nm in Al, and no less than 6 nm in Au. Thus the resolution promised by the focused probe of Ångstrom size is lost, even in a light element such as Al.

EDX is a useful technique for determining the composition of the sample. But it is intrinsically incompatible with analysis at atomic resolution. Chapter 7 discusses both EELS and EDX.

3.4.6 Other Techniques

There are many other signals available. Radiation in the visible or infrared can be observed, termed cathodoluminescence or CL. Near a semiconductor p-n junction, electron beam-induced current, EBIC, can be monitored. Both these outputs rely on the recombination of carriers produced by the inelastic scattering of the fast electron. The recombination lengths are long, usually hundreds of nm, or equal to the foil thickness, whichever is smaller (Yuan *et al.*, 1989). Copious Auger electrons as well as X-rays are emitted whenever there is an electronic transition, and these can be detected. In all of these cases, the signal fed into the spectrometer or detection system may originate far from the probe, and so cannot convey information at atomic resolution. Nevertheless, optical spectrometry produces much higher spectral resolution than can be achieved by electron spectrometry.

3.5 BEAM DAMAGE AND BEAM WRITING

As mentioned in section 1.5, usually, the main limitation of STEM is the damage to the specimen caused by the fast electron beam. A measure of the localised damage is the Gray, Gy, the energy deposited per unit mass in Joules per kilogram. The 100 kV beam suffers inelastic energy losses mostly by generating plasmons, each with energy around 20 eV, and with a mean free path for production of about 100 nm.

Thus one finds for a probe of 1 nA into 1 nm pixel a radiation absorbed dose of 10^{12} Gy. A lethal dose for humans is about 3 Gy. It should come as no surprise that analytical information is limited mainly by radiation damage. Despite this immense dose, the temperature rise in the probe is very small, less than one degree, because the probe itself is so small. What the plasmons do in covalently bonded insulators is produce electron-hole pairs and phonons. The electron-hole pairs (essentially ionisation damage) destroy the covalent bonding. The electron beam also imparts momentum directly to the ions, producing displacement damage – a vacancy and an interstitial. Particularly for light elements, and for metals, this is the critical damage process. In ionic compounds, multiple ionisation of the anion can occur, resulting in lattice breakdown and radiolysis, in which the anions form a gas bubble, surrounded by a cation-rich shell. In general, ionisation damage becomes more severe the lower the accelerating voltage of the beam, and the less electrically conducting the sample: it is at its worst in aliphatic organic materials. However displacement damage becomes more severe the higher the accelerating voltage and the smaller the atomic mass of the sample. It is at its worst in hydrogen-containing materials. It is probable that there is an optimum accelerating voltage for any problem.

In any practical problem these various modes of damage are certain to occur, and it is always necessary to check whether they affect the observations. This can only be done by experiments designed to see the consequences of damage and how much time it takes to develop under the experimental conditions. The most effective avoidance of radiation damage was achieved by Muller *et al.* (1996) who in an investigation of boron at grain boundaries in a nickel-aluminium alloy coated the sample with a thin layer of amorphous carbon. They demonstrated that this prevented emission of boron atoms from the foil by displacement damage and enabled them to acquire very informative EELS.

One very dramatic consequence of the damage is the prospect of beam writing to record text or to produce very fine conducting elements for electronics (Broers, 1978), or other artefacts such as electron diffraction gratings and lenses (Ito *et al.*, 1998). STEM is ideally configured to make use of beam damage in this way. So far, the nanostructures produced are regarded as curiosities, but they may play a useful role in the future.

3.6 CORRECTION OF SPHERICAL ABERRATION

The objective lens in CTEM and the probe-forming lens in STEM have traditionally been round iron-cored magnetic lenses. It is relatively easy to make stable magnetic fields because the inductance is large and smoothes out fluctuations in current. As such lenses automatically confer cylindrical symmetry on the electron beam, they can be made with a very small bore, and they can be very strong. A millimetre focal length enables very high magnification at practical object distances of several centimetres. However, such lenses focus the beam by a second-order effect: the axial magnetic field first causes cyclotron rotation of the electrons, and then convergence to the axis. The focal length is inversely proportional to the square of the magnetic field, and the lens is always a convergent lens. It was realised early on by Scherzer (1936) that such lenses must suffer from spherical aberration: they are only first-order focusing devices, so that the further a ray departs from axial, the closer to the lens it is brought back to the axis. In a later paper, Scherzer (1947) suggested several ways around the problem, including the use of the famous 'Scherzer defocus' which balances out spherical aberration by underfocusing the lens. This simple technique enables optimum quality phase-contrast, but of course the resolution achievable is much worse than it should be, about fifty times the wavelength of a 100 kV electron.

As discussed in Chapter 4, one route to better focusing is to try to use magnetic fields perpendicular to the optic axis: dipoles, quadrupoles, etc. which directly bend the beam. Other routes involve time-dependent excitation of lenses, or the introduction of electrostatic components in the train of focusing elements. Krivanek et al. (2008) briefly outlined attempts along these lines in the 1960s which worked in principle but which proved too complex to manipulate in a practical microscope. Krivanek recognised that advent of computer control transformed the situation, and that it should be possible to design controllable multipole arrays to eliminate the aberrations. Thanks to twenty-first century computer technology, electron microscopes had arrived at the stage achieved by optical microscopes in the mid-nineteenth century!

There are now available two classes of aberration-corrected microscopes: those derived from STEM, sometimes called dedicated STEM, and those derived from CTEM, but often incorporating extra postspecimen lenses to enable EELS. These are sometimes designated TEM/STEM or even more cryptically (S)TEM. Generally speaking, users of the latter class of instruments concentrate on the acquisition of images. Because the images are acquired in parallel (without scanning) they can combine both atomic resolution and a large field of view, typically micron-sized. Analysis is most easily carried out by EDX because the electron optics for EELS is complicated. In such microscopes, it is possible to see an image of the focussed probe. Users of dedicated

STEM concentrate on high-resolution images from smaller fields of view, typically 100 nm on an edge. It is not possible to form an image of the probe, but indirect methods are used to tune the microscope and assess the probe size. The great strength of dedicated STEM is the use of EELS to determine the chemical composition and nature of the chemical bonding on an atomic scale, and the ready availability of high-quality convergent beam diffraction patterns from the focussed probe. It is striking that at the time of writing, although the number of (S)TEM instruments far exceeds the number of dedicated STEM instruments, the number of papers using atomic resolution analysis in STEM far exceeds the number from all other instruments.

3.7 WHAT DOES THE FUTURE HOLD?

For the first time, electron microscopes are capable of genuine atomic resolution and simultaneous chemical analysis, albeit at the risk of intense radiation damage. There is a huge amount of work to be done to explore the novel world opening up before us. It is important to try to select problems where atomic structure actually affects macroscopic performance.

Any surface used to accelerate chemical reactions by catalytic action, any interface whose bonding is controlled by local composition, any biochemical process taking place in a localised cellular domain, any defect in solids responsible for chemical or optical activity: these are obvious subjects for study. Chapter 8 presents a selection of appropriate examples. However, it is always important to remember that at the present time atomic resolution is obtainable only in a high vacuum environment, and under intense electron bombardment.

As for future instrumentation, the domain of time-dependent reactions is hardly touched. It is also seems likely that microscopes which can select the operating voltage to minimise damage and optimise the desired information will be built and will enable better experiments to be performed. Meanwhile, there is a lot to be done with existing facilities.

REFERENCES

Ahn, C.C. and Krivanek, O.L. (1983) *EELS Atlas* (Gatan Inc., Warrendale, PA); available from Gatan (www.gatan.com).

Broers, A.N. (1978) *Electron Microscopy 1978*, J.M. Sturgess (ed), Vol. III, pp. 343–354, published by Microscopal Society of Canada; also Mochel, M.E.,

Humphreys, C.J., Eades, J.A., Mochel, J.M., and Petford, A.M. (1983) *Appl Phys. Lett.* 42, 392–394.

Brown, L.M. (1981) Scanning Transmission Electron Microscopy: Microanalysis for the Microelectronic Age, *J. Phys. F: Metal Physics*, 11, 1–26.

Cowley, J.M. (1969) Image Contrast in Transmission Scanning Electron Microscope, *Appl. Phys.Lett.* 15, 58.

Crewe, A.V. (1971) High Intensity Electron Sources and Scanning Electron Microscopy, in U. Valdrè (ed.) *Electron Microscopy in Material Science*, Academic Press, pp. 162–207. A separate account of the instrument can be found in Crewe, A. V., Wall, J., and Welter, L.M. (1968) A High Resolution Scanning Transmission Electron Microscope, *J. Appl. Phys.* 30, 5861. Atomic imaging was first reported by Crewe, A.V., Wall, J., and Langmore, J. (1970) Visibility of Single Atoms *Science* 168, 1338. It may interest readers to learn that Crewe is a graduate of Liverpool University, the major sponsoring university for SuperSTEM.

Egerton, R.F. (1996) *EELS in the Electron Microscope* (2nd edn), Plenum, N.Y.

Grigson, C.W.B. (1965) Improved Scanning Electron Diffraction System *Rev. Sci. Instruments*, 36, 1587.

Isaacson, M. and Johnson, D. (1975) The Microanalysis of Light Elements Using Transmitted Energy Loss Electrons, *Ultramicroscopy* 1, 33–52.

Ito, Y. Bleloch, A.L. and Brown, L.M. (1998) Nanofabrication of Solid-State Fresnel Lenses for Electron Optics, *Nature*, 394, 49–52.

Krivanek, O.L., Dellby, N., Keyse, R.J., Murfitt, M.F., Own, C.S and Szilagyi, Z.S. (2008) Advances in Aberration-Corrected Scanning Transmission Electron Microscopy and Electron Energy-Loss Spectroscopy, *Adv. Imag. Elec. Phys.*, 153, 121–160.

Leapman, R.D. and Cosslett, V.E. (1976) Extended Fine Structure Above the X-ray Edge in Electron Energy Loss Spectra, *Phil Mag* 331; *J. Phys. D.: Appl. Phys.*, 9 L29.

Metherell, A.J.F. and Whelan, M.J. (1965) RESEARCH NOTES: The Characteristics of the Möllenstedt Velocity Analyzer, *Brit. J. Appl. Phys.*, 16, 1038–1040; also Cundy, S.L., Metherell, A.J.F., Unwin, P.N.T. and Nicholson, R.B. (1968) Studies of Segregation and the Initial Stages of Precipitation at Grain Boundaries in an Aluminium 7wt.% Magnesium Alloy with an Energy Analysis Electron Microscope, *Proc. Roy. Soc.*, A307, 267–281.

Muller, D.A., Subramanian, S., Batson, P.E., Silcox, J. and Sass, S.L. (1996) Structure, Chemistry and Bonding at Grain Boundaries in Ni3Al–I. The Role of Boron in Ductilizing Grain Boundaries, *Acta Materialia*, 44, 1637–1645; Muller, D.A. and Silcox, J. (1995) Radiation Damage of Ni3Al by 100keV Electrons, *Phil. Mag.* 71, 1375–1387.

Rodenburg, J.M. (2008) Ptychography and Related Diffractive Imaging Methods, *Advances in Imaging and Electron Optics*, 150, 87–184.

Ruska, E. (1986) According to Ernst Ruska, who with Gerd Binnig and Heinrich Rohrer, won the Nobel Prize in 1986. With Max Knoll, Ruska invented the electron microscope, in the early 1930s, whereas Binnig and Rohrer in the early 1980s invented the Scanning Tunneling Microscope (STM, not to be confused with STEM). See Ruska's autobiography: http://nobelprize.org/nobel_prizes/physics/laureates/1986/ruska-autobio.html

Ruthemann, G. (1941) Direct energy Losses of Fast Electrons by Transmission Through Thin Sheets, *Naturwissenschaften*, 29, 648 and (1942) Electron Retardation at X-ray Levels. *Naturwiss*. 30, 145; also Hillier, J. and Baker, R.F. (1944) Microanalysis By Means of Electrons, *J. Appl. Phys.*, 15, 663–675.

Scherzer, O. (1936) Über einige Fehler von Elektronenlinsen, *Zeits. für Physik*, 101, 593.

Scherzer, O. (1947) Sphärische und chromatische Korrektur von Elektronenlinsen, *Optik*, 2 114.

Treacy, M.M.J., Howie, A, and Pennycook, S.J. (1980) Z Contrast of Supported Catalyst Particles in the STEM, *Inst. Phys. Conf. Ser. No. 52* (EMAG 79), T. Mulvey (ed.), I.o.P. Publishing, Bristol, p. 261.

Yuan, J., Berger, S.D. and Brown, L.M. (1989) Thickness Dependence of Cathodoluminescence in Thin Films, *J. Phys.*: Condens. Matter 1 3253 doi: 10.1088/0953-8984/1/20/006.

4

Lens Aberrations: Diagnosis and Correction

Andrew Bleloch[1] and Quentin Ramasse[2]

[1] *School of Engineering, University of Liverpool, Liverpool, UK*
[2] *SuperSTEM Facility, STFC Daresbury Laboratories, Daresbury, Cheshire, UK*

4.1 INTRODUCTION

We will swap between the two approaches previously introduced in Chapter 2 to describe optics in this chapter; geometric optics and wave optics. Wave optics is the more complete description in which the progress of a wavefront through the optical system is calculated according to a wave equation. This would be Schrödinger's wave equation for electrons and Maxwell's equations for electromagnetic waves. In practice, generic properties of waves are common across many wave equations and can be used to describe the essential behaviour of a system that would range from light to electron waves or even sound waves without needing to invoke the specific wave equations. Geometric optics is an approximation where the limit of short wavelengths is taken. This results in a description in which straight lines or rays can be used to describe the propagation of points on a wavefront and is extremely useful in visualising and understanding the operation of complex optical systems.

Aberration-Corrected Analytical Transmission Electron Microscopy, First Edition.
Edited by Rik Brydson.
© 2011 John Wiley & Sons, Ltd. Published 2011 by John Wiley & Sons, Ltd.

The most general (but not necessarily the most useful) definition of a lens is an optical element that modifies a wavefront in a manner that varies across the wavefront. This would include round (rotationally symmetric) lenses and multipole lenses. Here we will concern ourselves with round lenses to begin with. An ideal round lens can be simply defined as an optical element that causes a ray to deviate by an amount proportional to the distance from the optic axis as shown in Figure 4.1. Since this focuses a plane wave to a point image, the effect is to cause an incident plane wavefront to become part of a spherical wave that would then come to a point focus in the limit of short wavelengths. It is the deviation from this simple geometry that is the subject of the rest of this chapter and the drive for all of aberration-corrected microscopy.

Before embarking on our journey correcting aberrations, a somewhat deeper understanding of the operation of round magnetic lenses as used in electron microscopes is helpful, though not essential. Magnetic lenses cause electrons to change direction by the Lorentz Force, F, as seen in Chapter 2 (this is the same as equation 2.3):

$$F = -e(E + v \times B) \tag{4.1}$$

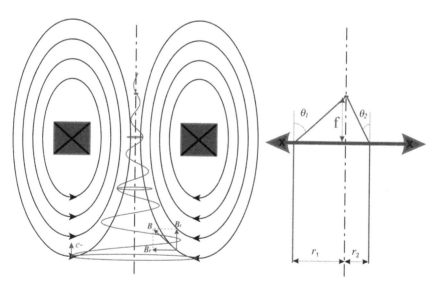

Figure 4.1   Ray diagram of a round magnetic lens and perspective of field lines. While the component B_t of the magnetic field has the primary effect of sending electrons travelling down the optic axis into a spiral, the focusing effect is due to the variation in B_r, the radial component of the magnetic field. In the geometric description, an ideal electromagnetic round lens deviates rays by an angle θ_i proportionally to their distance r_i the optic axis.

where E is the electrostatic field (which we assume to be zero for purely magnetic lenses), B is the magnetic field and v the velocity of the fast electron with charge $-e$. The vector cross-product means that the force due to the magnetic field is at right angles to the field lines. In other words, if the electrons are travelling parallel to the field lines there is no force.

The primary action of a round magnetic lens as shown in Figure 4.1 is therefore not to focus but to cause rays at an angle to the optic axis to spiral around the optic axis. Put another way, a constant field does not focus but a field that rises and falls in the z-direction does cause focusing. Unfortunately, in a round lens, the focusing effect of the field grows too rapidly with distance from the axis and the high angle rays are focused too strongly. Worse still, the so-called magnetic scalar potential (a quantity whose gradient gives the field lines) must obey Laplace's equation in free space and therefore, no matter what is done to manipulate the field lines, if the lens remains round then there will always be significant aberrations. Scherzer in 1936 proved that any electron optical system will always suffer from spherical aberration and chromatic aberration if simultaneously:

a. The optical system is rotationally symmetric.
b. The system produces a real image of the object.
c. The fields of the system do not vary with time.
d. There is no charge present on the electron-optical axis.

Although in principle breaking any one of these four conditions could allow spherical and chromatic aberrations to be corrected, in practice, the most success has been obtained by breaking rotational symmetry and this approach will take up most of this chapter. Clearly a real image is required but the last two conditions do deserve some discussion. Time-varying fields are used in practice in particle accelerators where, as a pulsed beam moves around the beam ring, at certain places in the ring the pulse is accelerated across a time-varying potential such that the slower particles arrive at a time when they get a bigger acceleration and the faster particles a smaller acceleration. This corrects chromatic aberration. The challenge that has not yet been overcome in order to apply this technique in the electron microscope is to obtain an electron beam with sufficient current and brightness that is pulsed with sufficiently short pulses (a $100\,kV$ electron is in the field of a typical electron lens for 20 picoseconds). The final of Scherzer's four conditions, introducing a charge on the electron-optical axis, is the second most pursued mechanism for correction. Two broad approaches have been attempted. The

first is to literally generate a beam of charged particles with an appropriate spatially-distributed charge density (increasing quadratically from
the optic axis) and then to pass the electron beam through this charge.
This approach has not yielded any significant useful optical improvement. Another approach, which relies on having a charge on the axis in
a slightly less obvious way, is to have a thin conducting foil (or mesh)
in the beam path. Being able to define an equipotential on the axis in
this way allows the aberrations to be corrected. Whilst this approach
has technical challenges, it is still being pursued and may yet prove
practically useful (see references in Hawkes, 2009).

Returning to the current prevalent technology based on breaking the
rotational symmetry, for most of this chapter, we will consider a minimal
optical system of a single lens that focuses an ideal point source to a
probe. The performance of most microscopes is largely determined by
the objective lens as this usually has the largest product of field of view
and angle of any lens in the column so treating a single lens is not
unreasonable. However, almost tautologically, once the aberrations in
this lens are corrected, it then becomes necessary to keep track of the
'budget' of aberration in the rest of the column. This is particularly true
of incoherent aberrations discussed in the next paragraph.

Ultimately an electron-optical system is designed to achieve a certain
resolution at a certain electron energy. To revise what we saw in
Chapter 2, the aberrations that constrain that resolution are usefully
divided into categories according to the means required to mitigate
their effects. The **geometric aberrations**, of which spherical aberration
considered by Scherzer is an example, are caused by the geometry of
the lens electromagnetic field. They are also called coherent aberrations
in that they distort the wavefront but retain the coherence. **Diffraction**
is not so much an aberration but a fundamental property of waves
that imposes an ultimate limit on resolution that simply depends on the
maximum aperture angle and the wavelength of the radiation. Although
the ray-optical description is useful and intuitive, it does not include
the limit to resolution imposed by diffraction. **Incoherent aberrations**,
of which chromatic aberration is often the most important, also limit
resolution. Incoherent aberrations also distort the wavefront but do not
distort all the waves in the same way – axial chromatic aberration, for
example, moves the focal point of the wave depending on the energy of
that wave. The information in the image is thus blurred by the addition
of these different contributions that cannot be unscrambled and are thus
incoherent. We could also think of mechanical and electrical instabilities
as being a kind of incoherent aberration.

4.2 GEOMETRIC LENS ABERRATIONS AND THEIR CLASSIFICATION

As indicated earlier, a perfect lens takes a plane wave and produces a perfect spherical wave that comes to a point focus. In the ray-optical description, the focus is indeed a point because the wavelength in this limit is taken to be infinitely small. For waves of finite wavelength, the size of the point is limited by diffraction. In describing the optics of an electron microscope, it is useful to separate the first-order (or Gaussian) trajectories from the practical aberrated trajectories. The first-order trajectories are the ray paths that depend only on simple focusing (in other words perfect lenses) and give the positions and magnifications of the image at the relevant planes as the rays progress down the column. It is called first order because ideal focusing is assumed where the ray is deviated linearly with its distance from the optic axis. Although the electrons of course do not care what we call their path through the column, it is conceptually helpful (and almost universally adopted) to consider the first-order trajectories as ideal and then to think of the deviation from these ideal trajectories as aberrations. This was introduced by Born and Wolf (2002), amongst others. This also means that first-order focusing has a different status and a more fundamental effect than the higher order aberrations. In other words, first describe the first-order trajectories and then the consequent aberrations. Here we are considering one lens and so the first-order trajectory is simply ideal focusing i.e. the spherical wave.

The wave aberration function (called χ here) is defined as the phase difference between this perfect spherical wave and the actual wavefront for a given lens as shown in Figure 4.2. In other words, *this phase difference is simply W, the error in the wavefront measured as a distance, multiplied by* $2\pi/\lambda$ (λ is the wavelength of the electron wave):

$$\chi = \frac{2\pi}{\lambda} W \qquad (4.2)$$

This two-dimensional surface is usually expressed as a function of the angle to the optic axis (θ in the notation below) and azimuthal angle around the optic axis (φ in the notation below). Obviously, for a perfect lens, $\chi(\theta, \varphi)$ would be zero everywhere but in practice it is a function that, when a microscope is in the best imaging condition, is close to zero near the optic axis but then increases increasingly rapidly as high-order aberrations come to dominate away from the optic axis.

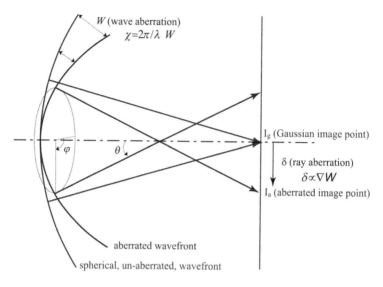

Figure 4.2   Wave aberration and ray aberration. The wave aberration W measures the distance by which an aberrated wavefront deviates from the ideal spherical case. The corresponding phase difference is called the aberration function, χ. The shift of the Gaussian image point due to the aberrations is called the ray aberration. The ray aberration δ is proportional to the gradient of the wave aberration W.

It is straightforward to calculate the STEM probe using this aberration function. For an ideal lens, the probe is simply the Airy disc that is the Fourier transform of the uniformly illuminated aperture in the back focal plane of the lens (see the Notation Appendix and Chapter 2 for a definition of the Fourier Transform and how it applies to lenses). To include aberrations, the probe function $\Psi_p(R)$ is therefore calculated as the Inverse Fourier Transform of the aperture function $A(\mathbf{K})$ in the back-focal plane, which includes the phase change $\exp(-i\chi(\mathbf{K}))$ due to aberrations and an aperture, which cuts off the wave at a particular angle, can also be included so that:

$$\Psi_P(R) = \int A(\mathbf{K}) \exp(-i\mathbf{R} \cdot \mathbf{K}) \, d\mathbf{K}$$

$$= \int H(\mathbf{K}) \exp(-i\chi(\mathbf{K})) \exp(-i\mathbf{R} \cdot \mathbf{K}) \, d\mathbf{K} \qquad (4.3)$$

where H is the aperture shape (that has a value of one inside the aperture and zero outside) and K is a two dimensional vector in the back focal plane of the probe-forming lens. Note here the aperture function $A(\mathbf{K})$ is defined as $H(\mathbf{K}) \exp(-i\chi(\mathbf{K}))$. One could choose a number of ways to fit a mathematical function to this surface but it turns out that some expansions are easier to use than others. In particular if the lens aberrations tend to zero for rays very close to the optic axis, which is usually the case for practical systems, the following expansion is useful (Krivanek *et al.*, 1999, and Hawkes and Kasper, 1996):

$$\chi(\theta, \varphi) = cst + \theta\{C_{01a}\cos(\varphi) + C_{01b}\sin(\varphi)\}$$

$$+ \frac{\theta^2}{2}\{C_{10} + C_{12a}\cos(2\varphi) + C_{12b}\sin(2\varphi)\}$$

$$+ \frac{\theta^3}{3}\{C_{23a}\cos(3\varphi) + C_{23b}\sin(3\varphi) + C_{21a}\cos(\varphi) + C_{21b}\sin(\varphi)\}$$

$$+ \frac{\theta^4}{4}\{C_{30} + C_{34a}\cos(4\varphi) + C_{34b}\sin(4\varphi) + C_{32a}\cos(2\varphi)$$

$$+ C_{32b}\sin(2\varphi)\}$$

$$+ \frac{\theta^5}{5}\{C_{45a}\cos(5\varphi) + C_{45b}\sin(5\varphi) + C_{43a}\cos(3\varphi)$$

$$+ C_{43b}\sin(3\varphi) + C_{41a}\cos(\varphi) + C_{41b}\sin(\varphi)\}$$

$$+ \frac{\theta^6}{6}\{C_{50} + C_{56a}\cos(6\varphi) + C_{56b}\sin(6\varphi) + C_{54a}\cos(4\varphi)$$

$$+ C_{54b}\sin(4\varphi) + C_{52a}\cos(2\varphi) + C_{52b}\sin(2\varphi)\}$$

$$+ \ldots \tag{4.4}$$

Apart from the initial constant term (*cst*), each term in the expansion is of the form:

$$\frac{\theta^{N+1}}{N+1}\{C_{NSa}\cos(S\varphi) + C_{NSb}\sin(S\varphi)\} \quad \text{or} \quad \frac{\theta^{N+1}}{N+1}\{C_{NS}\} \tag{4.5}$$

Although this expression may look intimidating at first, one can simply think of it as a simple polynomial expansion in polar coordinates of χ, a smooth, continuous two-dimensional surface, akin to a Taylor expansion that would have had its terms re-arranged and re-grouped for convenience. We will call the numerical coefficients C_{NS} that figure in front of the various polynomial groups the **aberration coefficients**.

Notice that if N is odd then S is even and vice versa. This is because of the symmetry: if N is odd then $\theta^{N+1} = (-\theta)^{N+1}$ and if N is even then $\theta^{N+1} = -(-\theta)^{N+1}$. Also, with this constraint of being even or odd, for each N, S takes values from 0 to $N+1$. When N is odd, then there is an $S = 0$ which is a round aberration with no azimuthal dependence (no sine or cosine oscillation in φ) – these are the most familiar aberrations like defocus and spherical aberration. In summary:

N odd ($N + 1$ even)	N even ($N + 1$ odd)
$S = 0, 2, 4 \ldots N + 1$	$S = 1, 3, 5 \ldots N + 1$

In equation (4.5), for non-round aberrations ($S \neq 0$) the C_{NSa} and C_{NSb} coefficients represent the projection of the overall aberration C_{NS} along two orthogonal axes (the sine and cosine projections in trigonometric terms), so the squared total amplitude of the aberration C_{NS} can be written as:

$$C_{NS}^2 = C_{NSa}^2 + C_{NSb}^2 \qquad (4.6)$$

Historically, aberrations have been classified according to the power law dependence of the *deviation of an off-axis ray* from the ideal Gaussian focus. This ray deviation δ is proportional to the gradient of the aberration function given in equation (4.4) above, as shown in Figure 4.2:

$$\delta \propto \nabla W \qquad (4.7)$$

(Note that this is not the same as the gradient of the wavefront – the aberration function is related to the error in the wavefront, W). Looking at equation (4.4), term by term the radial dependence of this gradient will look like:

$$\theta^N \{ C_{NSa} \cos(S\varphi) + C_{NSb} \sin(S\varphi) \} \qquad (4.8)$$

This historical description of the aberrations in terms of ray optics is why in equation (4.4) the aberration function has a radial dependence that is one order higher than the name of the aberration (third order aberrations have an θ^4 radial dependence in χ). Table 4.1 lists the different aberrations with the lower orders conventionally named; a similar table is also given in Appendix 1 with a slightly different nomenclature where the order is listed in terms of the wavevector, K, which is directly related to angle with the optic axis, θ.

The notation given here is that used by Krivanek *et al.* (1999) and hence often by users of Nion Co. aberration correctors. There are a number of alternative notations but it should be emphasised that they are all slightly different methods of expressing exactly the same information. Appendix 1 provides a translation between the different notation systems. Figure 4.3 illustrates the shape of a number of the

Table 4.1 The name, the order (of both the ray deviation (N) and the wavefront) and azimuthal symmetry (S) of the common aberration coefficients.

Aberration coefficient	Radial Order		Name	Azimuthal Symmetry
	Ray	Wave		
C_{01}	0	1	Image Shift	1
C_{10}	1	2	Defocus	0 (Round)
C_{12}			Twofold Astigmatism	2
C_{21}	2	3	Coma	1
C_{23}			Threefold astigmatism	3
C_{30}			Spherical aberration	0
C_{32}	3	4	Twofold astigmatism of C_s (or Third order twofold astigmatism)	2
C_{34}			Fourfold astigmatism of C_s	4
C_{41}			Fourth order coma	1
C_{43}	4	5	Fourth order threefold astigmatism	3
C_{45}			Fivefold astigmatism	5
C_{50}			Fifth order spherical aberration	0
C_{52}	5	6	Twofold astigmatism of C_5	2
C_{54}			Fourfold astigmatism of C_5	4
C_{56}			Sixfold astigmatism of C_5	6

important terms that contribute to the aberration function. Again, this plethora of aberrations could appear daunting but it is important to stress that these merely represent an extension of a nomenclature well-known to all microscopists. Looking more closely in Figure 4.4 at the influence that two of the lower order shapes have on an incoming wavefront, one recognises immediately that C_{12} (top of the diagram) will bring rays travelling at different azimuths to a different focus point – this is simply astigmatism, while C_{21} (bottom of the diagram) will displace radially the centres of concentric circles, thus distorting a round beam into a characteristic comet shape (with a 30° cone angle) – i.e. coma. Finally, another useful property of the aberrations to keep in mind is the remarkable similarity between terms with the same azimuthal symmetry order. C_{12}, C_{32} and C_{52} have essentially the same shape, as have C_{23} and C_{43} or C_{34} and C_{54}, and these groups of aberrations will therefore distort the incoming electron beam in a very similar way. The difference between those terms, however, is how steep the surfaces are further away from the optic axis (the 'centre' of the two-dimensional surface): thus C_{32} appears steeper that C_{12} in Figure 4.3; in other words, at low angles the lower-order aberrations dominate, which is just another way of saying that polynomials of a smaller order dominate close to zero.

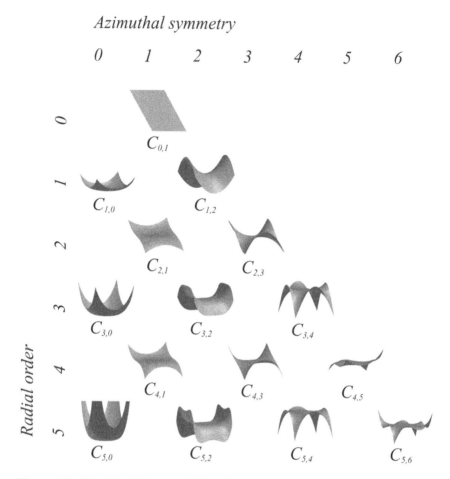

Figure 4.3   Graphical glossary of the aberration coefficients. Three-dimensional rendering of the distortions introduced in the wavefront by all aberrations up to fifth order. The aberration coefficients C_{NS}. can be classified by their radial order N and their azimuthal symmetry S. Notice how similar aberrations of identical azimuthal symmetry are.

The aberrations discussed so far are, so-called, axial aberrations. They limit the resolution on the optic axis of the lens as well as off-axis. To be general we should include aberrations which are zero on the optic axis but increase away from the optic axis. This introduces significant complications because we now need to describe, say, off-axial x-astigmatism that is now itself a function of x and y. It should also be mentioned that, in practice, because the signal in STEM images is typically small compared with the reciprocity equivalent CTEM images, the signal-to-noise

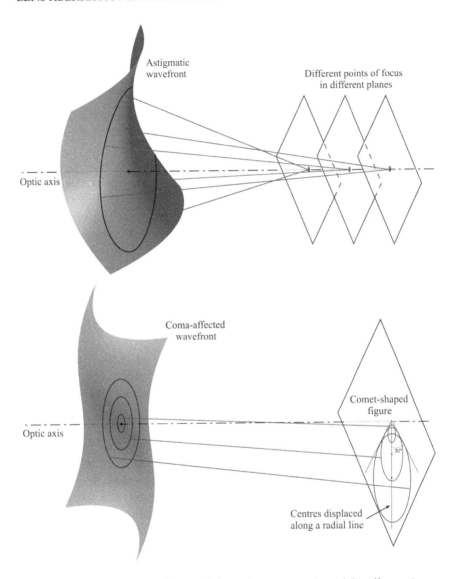

Figure 4.4 Astigmatism and coma. Schematic representation of the effect astigmatism (top) and coma (bottom) have on the beam. While an astigmatic wavefront focuses rays travelling at different azimuths onto different image planes, a coma-affected wavefront images a disc into a comet shape.

leads to relatively small fields of view. Put another way, we need a certain dwell time to collect sufficient signal – large fields of view therefore take an unfeasibly long time in practice. This has the consequence that off-axial aberrations do not often constrain practical imaging in STEM. The reader is referred to Hawkes and Kasper (1996) as well as recent papers for a discussion of these aberrations and their notation.

Spherical aberration has played a central role in electron microscopy. For five decades, it was the engineering constraint. It is not too much of an exaggeration to say that once the operating voltage was decided and the objective lens was designed, all of the rest of the microscope followed, including column size, sample holders, available tilt, required stability, aberration budget for the projection system, etc. This central role has led to instrumental resolution being defined in terms of only the spherical aberration in many electron microscopy books. The previous section hopefully cuts this spherical beast down to size and puts all aberrations on a more equal footing, providing a healthier perspective.

4.3 SPHERICAL ABERRATION-CORRECTORS

Breaking the first of Scherzer's assumptions that lead to inevitable spherical aberrations, (i.e. breaking rotational symmetry) is the practicable method that is the basis of the two currently commercially available technologies. This is achieved using multipole lenses, first outlined in section 2.4. Multipole lenses come in a series with 2, 4, 6, 8 . . . poles alternating North and South around the optic axis as shown in Figure 4.5. These are named dipole, quadrupole, sextupole and octupole respectively. (A sextupole is often called a hexapole which may be easier to pronounce but this red rag to a pedant introduces one Greek-based number in a set of otherwise Latin-based numbers!).

The most important difference between multipole lenses and round lenses is that in the former the field lines are mostly perpendicular to the beam while, as we saw earlier in this chapter and in Chapter 2, in the latter they are largely parallel to the beam. The way the Lorentz Force operates means that, in a multipole lens, the primary effect of the magnetic field is to achieve the desired deflection of the rays. This is in contrast to round lenses where the large axial component of the field has only the effect of causing the electrons trajectories to rotate. Multipole lenses therefore require fields that are an order of magnitude smaller than round lenses. The current-carrying solenoid coils are therefore smaller and the heat produced often insignificant. They also do not rotate the image as the strength of the lens is altered.

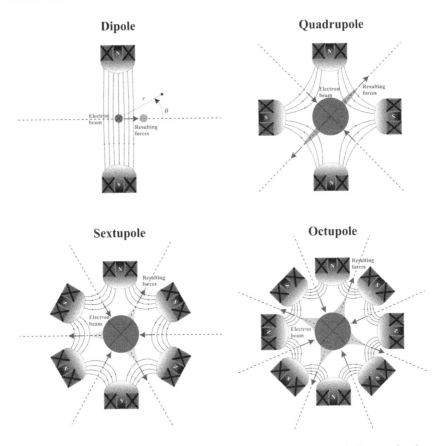

Figure 4.5 Multipole lenses. Schematic representation of multipole lenses of orders 2, 4, 6 and 8, with an impression of the corresponding field lines. Solving for the scalar potential at the polar coordinates (r, θ), the resulting forces on an electron beam travelling down the optic axis (into the page) have the following primary effect: a dipole shifts the beam, a quadrupole focuses the beam into an line, a sextupole imparts a triangular distortion to the beam and finally an octupole imparts a square distortion to the beam.

To calculate the effect that a multipole lens has on an electron trajectory, we need to know the magnetic field and to use the Lorentz force. The field in turn can be calculated from the magnetic scalar potential, which obeys Laplace's equation in free space. If the magnetic poles have an ideal shape then the solution to Laplace's equation for a (long) multipole lens consisting of $2N$ poles can usefully be expressed in the following form:

$$\Phi(r, \theta) = r^N\{p_N \cos(N\theta) + q_N \sin(N\theta)\} \qquad (4.9)$$

where the definitions of r and θ are shown in Figure 4.5 (see a standard electromagnetism textbook such as Jackson (1975), for a full justification of equation 4.9). The magnetic field is equal to the gradient of this magnetic scalar potential Φ. Notice how similar in form this equation is to equation (4.8) albeit with a different variable notation. Indeed a correctly orientated quadrupole will directly correct simple astigmatism. This is true for most of the aberrations except unfortunately for the dominant round aberrations. Cunning is therefore required to use a combination of the field shapes based on equation (4.9) to have the required effect on the beam. It is not possible to create a single field distribution that will achieve this in one plane (otherwise Scherzer's theorem would not hold) so it is necessary to build up the corrections in more than one plane.

The radial dependence of the fields associated with the scalar potential of equation (4.9) gives a direct indication of how to proceed. The quadrupole ($N = 2$) has a field proportional to the distance from the axis, which corresponds to a simple focusing effect albeit focusing in one azimuthal direction and simultaneously defocusing in the perpendicular direction, as shown in Figure 4.5. A quadrupole is therefore often described as producing a 'line focus'. The octupole on the other hand has a field that varies as r^3, which acts directly on third order aberrations. This lead Scherzer (1947) to propose the forerunner of what became the quadrupole-octupole corrector. This concept after many trials by many researchers was finally successfully implemented in an SEM by Zach and Haider (1995) and at atomic resolution in the STEM by Krivanek et al. (1999). The operation of this corrector is described in the next section.

The other design of corrector, the hexapole corrector, relies on a slightly more subtle effect. All the above assumes that the multipole lenses are thin. In other words they act in one plane. A long multipole element has an additional second order effect that can be understood by thinking of several thin weak multipoles (say quadrupoles) one after the other to make up the effect of one strong thin element. If a round beam enters the first quadrupole then a slightly elliptical beam will enter the second. The rays that are further from the axis will experience a stronger field than they would have, had the beam been round. This additional effect causes the beam to deviate from the axis with a rate that is one power higher in the radial direction. That is, a long quadrupole will still produce an ellipse but there will be a quadratic dependence on the strength of the effect in addition to the linear effect expected of a quadrupole. The hexapole corrector uses this principle by having two long hexapoles such that the second cancels out the first-order effect of the first but leaves twice the second-order effect. The first-order effect of a hexapole is quadratic in the radial direction and three-fold azimuthally but the second-order effect is cubic in the radial direction and uniform azimuthally i.e. a straightforward negative spherical aberration (see Table 4.1).

4.3.1 Quadrupole-Octupole Corrector

The quadrupole-octupole (QO) corrector traces its origin to a proposal by Scherzer (1947). The modern version shown schematically in Figure 4.6 consists in essence of four quadrupoles and three octupoles. As pointed out above, the field strength, and hence the deflection of an electron travelling through an octupole, varies cubically with distance from the axis. Unfortunately, although this does directly produce negative spherical aberration along two lines, it produces positive spherical aberration in between these two axes. The solution is to use a series of quadrupoles to squeeze the beam to (or close to) a line focus and put this highly elliptical beam along one of the directions with negative spherical aberration. In practice this is achieved with the arrangement of four quadrupoles shown in Figure 4.6. If the octupoles are turned off, the four quadrupoles act as a round lens and indeed this configuration of four quadrupoles can be used in place of a round lens. What is important for the correction of spherical aberration is that the round beam is extremely elliptical at planes O2 and O6 (the first and third octupoles, named here octupole 2 and octupole 6 because they are placed in the second and sixth optical layers of the corrector) in the x and y directions respectively (see Figure 4.6). The two octupoles at these planes can be adjusted to correct the spherical aberration in x and y respectively. These octupoles unfortunately introduce a certain amount of four-fold astigmatism (because the beam is not a perfect line focus) and so a third octupole acting on a round beam and of opposite sign to O2 and O6 is required to correct for this effect. This could be placed before or after the corrector but placing it in the middle, at the fourth optical layer, has some practical advantages. In summary, the progress of the beam through the corrector is as follows:

- Q1 (quadrupole 1) produces a crossover at Q3 in x and a highly elliptical beam at O2.
- O2 corrects spherical aberration in x and introduces some four-fold astigmatism.

- Q3 has little effect in x (because the ray is almost on axis) but produces a cross-over in y at Q5.
- O4, acting on an approximately round beam, corrects the four-fold astigmatism introduced by the combination of O2 and O6.
- Q5 has little effect in y but deflects the electrons in x so that a round but convergent (divergent) beam in x (y) arrives at Q7.
- O6 corrects spherical aberration in y and introduces some four-fold astigmatism;
- Q7 returns the beam to a round beam that now has acquired negative spherical aberration.

So the central problem is that, using multipole lenses, we cannot decouple the azimuthal symmetry from the radial dependence of the field – an octupole always has cubic radial dependence together with four-fold azimuthal symmetry whereas spherical correction requires round azimuthal symmetry (with the cubic radial dependence). We get around this by coupling aberrations together – extreme astigmatism giving almost a line focus, which allows the negative spherical aberration in one direction to dominate. One consequence of this approach is that we have introduced the most extreme astigmatism, which must be very accurately compensated. This is, in effect, the problem of so-called parasitic aberrations – aberrations inadvertently introduced by imperfections in the corrector. One final point before leaving the QO corrector: if we can make a lens with arbitrarily small aberrations out of multipoles, why do we need round lenses at all? The reason is instabilities in power supplies. The round lens performs (in STEM) a large demagnification of the instabilities in the multipoles, which allows the use of power supplies that do not have very expensive stability requirements.

4.3.2 Hexapole Corrector

The second-order effects of a long hexapole introduced in section 4.3 were first described by Hawkes in 1965 (see Hawkes, 2009) but their exploitation was not considered until Beck (1979) proposed a corrector design. This preliminary design was perfected by Rose (1981) and after successful implementation by Haider *et al.* in 1998 this has become the most widely used corrector geometry in TEM/STEM instruments.

As mentioned in section 4.3, it relies on two hexapoles, Hex1 and Hex2 that are rotated by 60° with respect to each other so that the first-order effect of the first is cancelled by the first-order effect of the second. However for this to happen to the desired precision, the effects of the

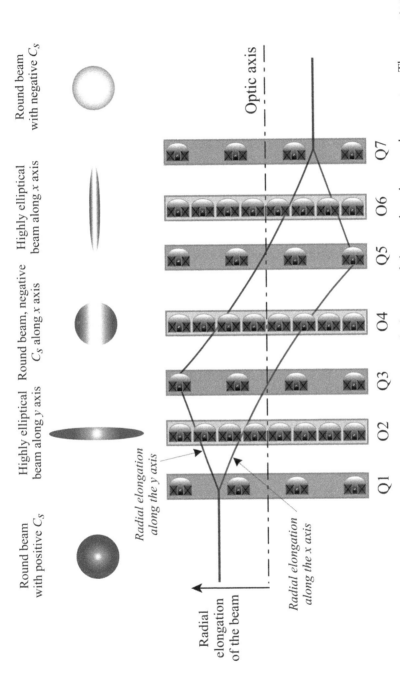

Figure 4.6 Quadrupole-Octupole corrector. Schematic and ray path diagram of the quadrupole-octupole corrector. The corrector consists of three octupoles (O2, O4 and O6) and four quadrupoles (Q1, Q3, Q5 and Q7). The radial elongation of the beam through the corrector is indicated along two orthogonal axes x and y (the line traces overlap when the beam is round).

two hexapoles must occur in equivalent optical planes. To achieve this, a pair of projector lenses (or transfer lenses) TL21 and TL22, is placed between the two hexapoles as shown in Figure 4.7, which also shows schematic trajectories through the corrector. As in the case of the QO corrector, the ideal simulated trajectories are straightforward. However, mechanical misalignments on the order of nanometers as well as drift mean that a number of small but important adjustments are needed in practice to achieve anything close to the theoretical performance. The operation of this corrector will be more completely described in Chapter 9 in the context of CTEM, which is equivalent, by reciprocity, to the STEM case. The major practical difference between CTEM and STEM operation is the aberration diagnosis method which will be described in section 4.5.

4.3.3 Parasitic Aberrations

Three factors are believed to have inhibited the development of aberration correctors. These are: stability of electronics, a detailed understanding of the role of parasitic aberrations and suitable algorithms and computational performance for quickly and accurately diagnosing aberrations. For the microscope user, the two important factors to understand are the parasitic aberrations and the diagnosis algorithms (the latter are addressed in section 4.5).

Parasitic aberrations are aberrations that have the same undesirable effect on the beam as all other aberrations but are so-called parasitic because they are inadvertently introduced by imperfections in the corrector itself. Put another way, it is possible to design an aberration-corrector on paper (or more usually now by calculating fields and trajectories in the computer) that works seemingly perfectly in simulation. However, building such a device with sufficiently small errors, in say the relative alignment of the components, is in practice impossible. Here we should quantify this error. In order to produce a desired resolution, aberrations of each order should produce a phase change across the aperture of less than $\pi/2$ (one quarter of a wavelength) of the electron wave. This means that, for example, astigmatism, which is described by the term in the aberration function (from equation 4.4):

$$\frac{\theta^2}{2}\{C_{12a}\cos(2\varphi) + C_{12b}\sin(2\varphi)\} \qquad (4.10)$$

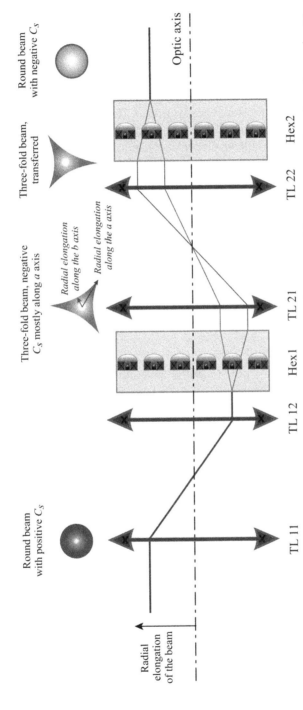

Figure 4.7 Hexapole corrector. Schematic and ray path diagram of a hexapole corrector. The corrector consists of two hexapoles (Hex1 and Hex2) and four round lenses arranged in two pairs of transfer doublets (TL11, and TL12; TL21 and TL22). The radial elongation of the beam through the corrector is indicated along two axes *a* and *b* as indicated (the line traces overlap when the beam is round).

will have a maximum value of $\frac{\theta^2}{2}\{C_{12}\}$, where $C_{12} = \sqrt{(C_{12a})^2 + (C_{12b})^2}$ is the magnitude of the overall astigmatism (see equation 4.6). This should be less than $1/4$, which means that the astigmatism needs to be less than roughly 2 nm in order to achieve 0.1 nm resolution in a 100 kV microscope (the resolution determines the aperture angle (α) through the diffraction limit and the accelerating voltage determines the electron wavelength). So, just as historically it was discovered that no round lens could be practically manufactured without astigmatism and hence stigmators have been the first simple aberration-corrector that we are all familiar with, understanding the diagnosis and correction of parasitic aberrations is crucial to realize the advantages of aberration-correction. This means that describing a corrector as a third-order corrector is short-hand for a corrector for ALL aberrations up to and including third-order.

Correction of each of the lower-order parasitic aberrations requires its own strategy:

• Astigmatism is adjusted by a rotatable quadrupole.
• Coma is adjusted by a dipole which alters the relative alignment of two (effectively) round lenses.
• Threefold-astigmatism is adjusted by a rotatable hexapole.

Each of these should be positioned in the column so that adjusting one does not alter the trajectories though the corrector and so inadvertently change higher order aberrations at the same time. Which brings us to the adjustment of third-order aberrations: this obviously can be achieved by altering the strength of the octupole lenses on a QO corrector and the hexapole elements in a hexapole corrector; however, from the brief descriptions of both the QO and hexapole correctors in the previous sections, it should also be apparent that adjusting the angular size of the beam in the corrector – all the while maintaining relative alignment and the relative strength of the different elements, will have a similar effect!

4.4 GETTING AROUND CHROMATIC ABERRATIONS

Given that aberrations initially up to third order (and subsequently up to fifth order in the latest generation of instruments) have been corrected, what consequences does this have for overall column design and achievable resolution? Chromatic aberration has become the limiting aberration in many cases and this is certainly true for operation at the lowest accelerating voltages. There are three approaches to limiting the effect of chromatic aberrations. The energy spread of the electrons contributing to the image can be reduced by filtering the electron beam in the gun, allowing a more monochromatic beam to impinge on the sample. This energy spread can also be reduced by filtering the electrons after they have passed through the sample (energy filtered transmission electron microscopy or EFTEM). This is more demanding because 0.1 eV in 300 keV requires a more precise filter than filtering before the electrons undergo most of their acceleration and hence energy gain. The third method actually corrects the optics. Correcting chromatic aberration works by first separating the electrons out according to their energy (say slower electrons further from the axis) and then putting these electrons through a field that corrects their trajectories (say increases the speed of these slower electrons). Given that the instrumentation for chromatic aberration-correction is in its infancy and the breadth of its applications still not fully realised, the reader is referred to the literature (see for example Kabius *et al.*, 2009 and Leary and Brydson, 2011).

4.5 DIAGNOSING LENS ABERRATIONS

Diagnosing the aberrations means measuring the aberration function, which in turn is the phase error in the wavefront that passes through the objective aperture. Because, as we have seen, the aberration function is usefully expanded as a polynomial (equation 4.4), this can analogously be estimated to any required order by measuring the value of the aberration function at a finite number of angles and then fitting a two-dimensional polynomial of the form of equation (4.4) to that data. Indeed, as we know, we can fit a one-dimensional polynomial of order n with at least $n + 1$ data points – the fewer the data points the more accurate they need to be to reproduce the underlying function. It turns out, however, to be rather difficult to measure the value of the aberration function itself directly, but, again, because the function is a polynomial, it is sufficient to measure its derivative (or even its second derivative) and to fit that. Fortunately, the slope (first derivative) and curvature (second derivative) are much easier to find reliably than the value of the aberration function itself. This is because the curvature of the wavefront (or in other words, its second derivative) is the local defocus of an image associated with an angle of illumination, while local image shifts are directly related to the wavefront surface slope (in other words, its first derivative). This is illustrated schematically in Figure 4.8. Obviously, in

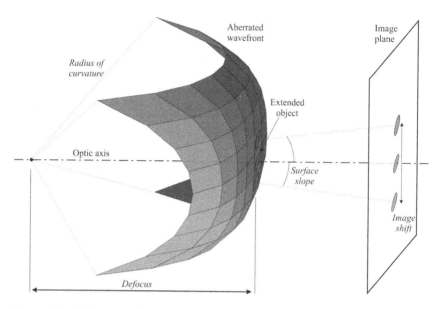

Figure 4.8 Effect at local angles of slope and curvature of the aberration function. Locally the slope of the aberration function can be related to a shift of the Gaussian point in the image plane, while its curvature is the same as the local defocus (approximating the surface locally by a sphere with a defined radius of curvature).

finding the second derivative, some information is lost – the constant term and the term proportional to theta – but fortunately these are an unimportant absolute phase (the constant term cst in the notation of equation 4.4) and an equally unimportant overall image shift ($C_{0,1}$ in the notation of equation 4.4).

Figure 4.9 shows schematically the operation of a wavefront detector called the Shack-Hartmann detector that is used to correct the wavefront in light-optical telescopes in real time to compensate for atmospheric distortions and imperfections in the mirror. Such a detector is intuitive to understand and helps to illustrate a principle on which most aberration diagnosis algorithms operate. The detector is based on dividing the incoming wavefront into subsections and using the fact that an aberrated wavefront will have a varying slope, which would cause the local sub-images to shift with respect to one-another. These measured shifts provide a method of finding the aberration function from multiple images of a single bright test object (the so-called 'guide star', in astronomy).

In the electron microscope we could think of a similar experiment where, if we had a regular array of point objects instead of an array of

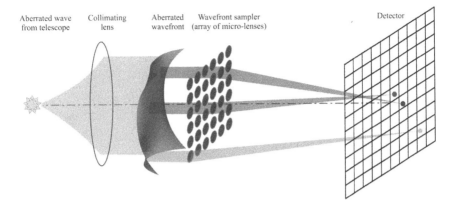

Figure 4.9 The Shack-Hartmann wavefront detector for telescopes. In a Shack-Hartmann detector the light incoming from the telescope is collimated into a plane wave before being divided by an array of micro-lenses. Any aberration in the wavefront will be immediately recognised at shifts in the detector plane.

microlenses, we could construct a method almost identical to the Shack-Hartmann detector. Here, we would measure the shifts in the positions of the projected images of those point objects instead of re-focusing part of the wave. Although such a method is used to align electron optical components that do not require the ultimate spatial resolution (the aperture that consists of an array of holes found in the Gatan GIF for example), unfortunately it is not practical for diagnosing aberrations here due to its coarseness. Therefore several slightly less direct techniques have been used to find the required slopes or curvatures of the aberration functions.

4.5.1 Image-based Methods

The simplest image-based method is very similar to the Shack-Hartman detector. If a number of detectors are placed in the bright field disc as shown in Figure 4.10 and the probe is scanned across the sample to form a bright field STEM image of the sample, then the image detected by each of the detectors will be shifted with respect to the central (untilted bright field) image. The measured shift δ at a 'sub'-detector placed at an angle θ and an azimuth φ to the optic axis (using the exact same variable notation and coordinates as for equation 4.4) is proportional to the gradient of the aberration function at that position, $\chi(\theta, \varphi)$. Figure 4.11 shows a montage of such images collected quasi-simultaneously by Krivanek, Dellby and Lupini during the development of the prototype

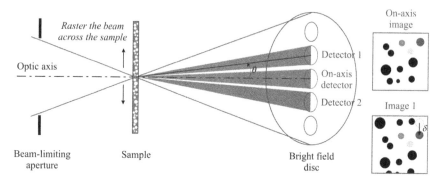

Figure 4.10 Shift tableau principle. Placing several detectors within the bright field disc allows to record simultaneously a series of bright field images (scanning the beam across the sample to generate the image serially). These images are effectively obtained with different beam tilts. The resulting shifts in the image can then directly be linked to the slope of the aberration function.

Nion corrector in Cambridge in the late 1990s. The relative shifts can be clearly seen and are indicated by exaggerated arrows on each of the images. The gradient of the aberration function can then be obtained from a least squares fit to the calibrated shift values thus obtained. This method has not been used by any suppliers of STEM correctors but this may be for legal reasons relating to intellectual property. If the images are collected simultaneously as set out here then sample drift would have no effect. However, one could imagine using one bright field detector and tilting the bright field disk to collect a series of images one after the other at different times. It should be obvious that the relative shifts so measured would be very susceptible to drift, particularly with the relatively slow image acquisition associated with STEM images.

Instead of evaluating the image shift, a variation on this method is to take the Fourier transform of the image at each angle and obtain the defocus and astigmatism at each of these angles from the Thon rings in the diffractogram. The reader is referred to a textbook on transmission electron microscopy (e.g. Williams and Carter, 2009) and descriptions of phase contrast in particular to understand how the defocus and astigmatism can be found from knowledge of the lens and the diffractogram of an amorphous sample. This is the exact reciprocity equivalent of the Zemlin tableau conventionally used to diagnose the aberrations in image- or TEM-corrected instruments (as opposed to the probe- or STEM-corrected instruments that we are discussing here) – see Chapter 9. While this 'STEM Zemlin tableau method' has the disadvantage of being somewhat more complicated

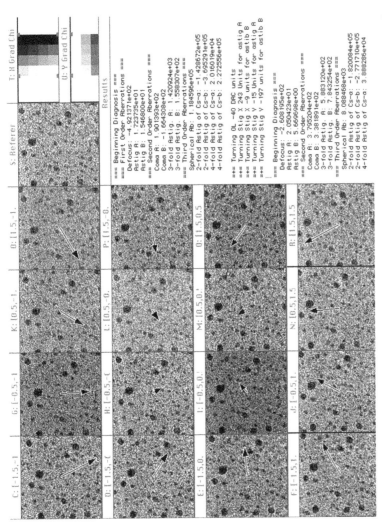

Figure 4.11 Shift tableau diagnosis in practice. Example of a shift tableau obtained by Krivanek, Dellby and Lupini during the development of the QO corrector at the Cavendish Laboratory, University of Cambridge in the late 1990s. A reference image (window S, top right) is compared to tilted bright field images obtained quasi-simultaneously, to yield an average shift and build a two dimensional map of the gradient of the aberration function (windows T and U, top right).

and sensitive to noise, it has the distinct advantage that it is relatively immune from spatial drift.

4.5.2 Ronchigram-based Methods

The ray optics of the formation of the Ronchigram is shown in Figure 4.12. As an imaging mode in electron microscopy it was first characterised and popularised by Cowley (Lin and Cowley, 1986) who also named it after Vasco Ronchi, an Italian physicist who had used an optically-equivalent mode to assess light optical lenses (Ronchi, 1964). There is an extensive literature on the Ronchigram, which can be variously described as a projection image, a shadow image, an inline hologram and an aberration map going back to the original paper by Lin and Cowley in 1986. Here we will exclusively concentrate on Ronchigrams of (sufficiently) amorphous samples for the purpose of diagnosing aberrations. The Ronchigram in Figure 4.12 was simulated by computer: in the model used, a typical amorphous sample (i.e. a random grey level pattern with a given feature size) was placed under the simulated electron beam at a moderate defocus (a few hundred nm), a value chosen to make the aberrations of the uncorrected lens obvious. Just as when placing an object in front of a torch (a typically diverging source of light) one projects a magnified shadow of it on a nearby wall, the amorphous speckle of the sample positioned in the electron beam is magnified in our detector plane. However, only the very centre of this

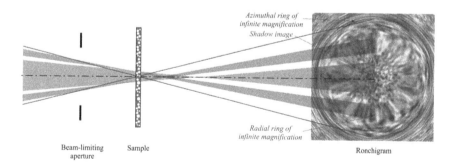

Figure 4.12 Ronchigram ray diagram. Ray optical diagram illustrating the formation of the Ronchigram. The simulated amorphous sample is placed at moderate underfocus: due to spherical aberration, distorted features called the rings of infinite magnification appear around the sample projection image. The Ronchigram was simulated with a defocus of −500 nm and for a spherical aberration of 3.1 mm).

image faithfully transfers the spatial information of the sample; this is known as the 'sweet spot' – an optimum (objective) aperture size for STEM imaging can be selected by matching the aperture diameter to the diameter of this sweet spot in the Ronchigram (see section 6.2.6). Distortions, here due to spherical aberration, are obvious outside of this small central patch. Two important 'distorted' features are the two, so-called, rings of infinite magnification. The easiest to explain is the azimuthal ring of infinite magnification, which appears at the angle at which the overfocus due to spherical aberration causes rays to be focused on the centre of the sample. In a case such as here where spherical aberration dominates, there will be an entire cone of such rays, due to the circular symmetry: therefore, a single point (a feature of dimension 'zero') at the sample will be magnified into a ring (a feature of dimension 'one'), hence the term infinite magnification. If the angle of this ring can be measured and the defocus is known it is, in principle, straightforward to determine the spherical aberration at least in this case where it is the dominant aberration. More generally, it should be obvious that other aberrations will influence the shape of the rings of infinite magnification: astigmatism, for instance, would cause the rings to be elliptical instead of round. While it is relatively simple in this case to describe the aberrations, a more mathematical treatment is needed to explain the general case, especially when aberrations of the same azimuthal symmetry but different radial order are present (for example, C_{32} would also make this ring elliptical). Similarly, the radial ring of infinite magnification arises when a pencil of adjacent rays forms a cross-over at the sample, thus imaging a single point into a line segment.

We have already seen in equation (4.7) that the deviation of a ray at the Gaussian focal plane is proportional to the gradient of the aberration function. We can extend this argument by looking at the far-field as the rays propagate away from the Gaussian plane as shown in Figure 4.13. The Ronchigram becomes, by construction, a map of the gradient of the aberration function. This is most easily seen when the only aberration is a large defocus. In this case, it is clear that the ray deviation varies linearly across the sample and so produces a magnified projected shadow-image of the sample in the Ronchigram (in the ray optics approximation of course). The magnification of this image is uniform because defocus has a linear gradient. Defining the magnification as the ratio between the projected size Σ at the detector plane of a feature of size σ at the sample, it can be straightforwardly calculated as the ratio of the distance between the Gaussian focus and the sample (Δf) and the Ronchigram camera length (L):

$$M = \frac{\Sigma}{\sigma} = \frac{L}{\Delta f} \qquad (4.11)$$

If other aberrations are present, then the rays do not all cross at a single Gaussian image point: instead, the apparent source varies across the Ronchigram as shown in Figure 4.13b, as though there were a different apparent defocus at every point. Just as in the previous case where a

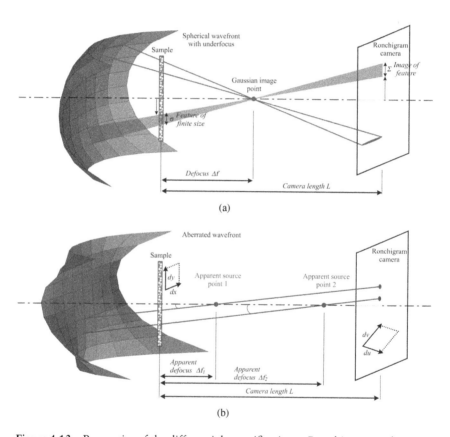

(a)

(b)

Figure 4.13 Ray optics of the differential magnification – Ronchigram analysis. (a) When only defocus is present, all rays going through the sample cross at a unique Gaussian point. The magnification of a feature of finite size σ into a feature of size Σ on the image plane can be expressed as a function of defocus Δf and camera length L. (b) Other aberrations distort the wavefront, giving each image point locally an apparent source and an apparent defocus. The local (differential) magnification is defined as the magnification of an infinitesimally small 2D feature (dx, dy) in the sample plane into (du, dv) in the image plane.

uniform magnification could be defined, these variations of the source point (or apparent defocus) can be interpreted as arising from a locally variable magnification. It turns out that the change in magnification between two points in the Ronchigram separated by an infinitesimal distance (in other words, the local or differential magnification) is proportional to the second derivative of the aberration function. Therefore, if we measured the local magnification in the Ronchigram plane at enough points, we could obtain a fit to the relevant second derivative of the aberration function polynomial. And since it is a polynomial, we can then integrate to recover the aberration function.

In practice, the local magnification is found at several points in the Ronchigram by moving the probe across the sample by a small (calibrated) amount in a particular direction (typically, two amounts dx and dy in orthogonal directions: see Figure 4.14). These probe shifts will naturally generate corresponding shifts (du and dv) in the Ronchigram. Because of the aberrations, the observed Ronchigram shift is not uniform and it is necessary to measure the distances (du and dv) at several observable points (or patches) in the Ronchigram (using cross-correlation). Naturally, the number of required patches depends on the order to which the aberration function is being fitted. In the

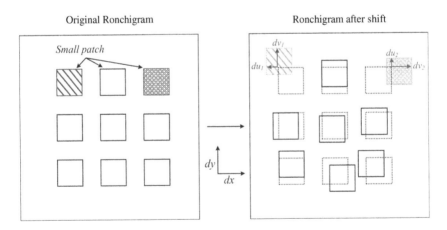

Figure 4.14 The Ronchigram analysis algorithm. After an initial Ronchigram is acquired (stationary probe in slight underfocus), the probe is moved by two calibrated amounts (dx, dy) in orthogonal directions. At the new probe position, a new Ronchigram is acquired. The gradient of the aberration function is estimated at several points across the Ronchigram by measuring the shifts (du_i, dv_i) of features within small patches. Note that each individual patch is a sub-shadow image, almost identical to a tilted bright field image (only the centre patch is untilted).

one-dimensional case, the local magnification would have simply been written as du/dx, but, because this is a two-dimensional problem, relating (dx, dy) to (du, dv) becomes a matrix equation. Each of these measurements therefore gives the local value for the differential magnification according to the equation:

$$(dx, dy) = \frac{1}{L}H(du, dv) \qquad (4.12)$$

where just as for equation (4.10) L is the Ronchigram camera length and H is a transformation matrix. In equation (4.11), the transformation factor was simply a uniform defocus value, so here, by analogy, H should be related to the local apparent defocus. In other words, and remembering the introduction to this section and Figure 4.8 in particular, H can be thought of as a local curvature field and it is indeed simply the second derivative (or 'Hessian') of the aberration function:

$$H = \begin{bmatrix} \dfrac{\partial^2 \chi}{du^2} & \dfrac{\partial^2 \chi}{du\,dv} \\ \dfrac{\partial^2 \chi}{dv\,du} & \dfrac{\partial^2 \chi}{dv^2} \end{bmatrix} \qquad (4.13)$$

It should be noted that although this procedure has been very successfully implemented by Nion Co., unless care is taken, errors in measurements can lead to large errors in the measured aberrations (the fitting is a so-called 'stiff' problem).

In a simpler but not so widely applicable method, we can deduce the local defocus by looking at a single Ronchigram of an amorphous sample, again, in point projection. Each individual patch (as those shown in Figure 4.14) used to evaluate the local differential magnification can almost be thought of as a bright field image in its own right, albeit recorded with a certain beam tilt, as long as the aberrations are not too severe so the 'shadow projection image' within the Ronchigram plane extends to large enough angular values. The geometry then becomes almost identical to the multiple bright field detector illustrated earlier in Figure 4.10. If we cut out those sub-images and then take their Fourier Transform, we can again deduce the defocus and astigmatism from the Thon rings in this so-called diffractogram. This is close to the reciprocity equivalent of the Zemlin tableau used to diagnose aberrations in the TEM (it is in fact performing all the tilts in the Zemlin tableau simultaneously). Although this is possible and is under investigation as a practical method, most microscopes do not have enough pixels in their Ronchigram camera to use this method.

As a final note, all the above methods rely on contrast in the Ronchigram of a thin amorphous film being dominated by scattering from a thin pencil of rays in the forward direction. In other words almost all the scattering must take place in the primary beam direction, which makes the Ronchigram of a thin amorphous film almost equivalent to a bright field image. In practice all these methods work and so this must be the case (at least at the medium to large defocus values used across all such methods) but this is not trivial to justify.

4.5.3 Precision Needed

As indicated in section 4.3.3, the precision with which the low-order aberrations need to be adjusted (a few nm for astigmatism for example) means that the diagnosis algorithms need to be sufficiently accurate. This is particularly challenging since the calculation of the aberrations from the data is rather susceptible to error – a small error in one of the measurements can cause an aberration like third-order astigmatism (C_{32}) to be partially assigned by the least-squares fit to first-order astigmatism (C_{12}) as they have the same symmetry and therefore distort the wavefront in similar ways. In practice it has been found by most users that the lowest-order aberrations (defocus and astigmatism) drift the fastest and are often best adjusted by eye (experience and cunning) and correspondingly the higher orders, particularly the third-order aberrations, are sufficiently stable over time to retain the required resolution.

4.6 FIFTH ORDER ABERRATION-CORRECTION

The so-called Scherzer optimal defocus was mentioned in Chapter 3, section 3.6: it is that (negative) defocus which optimally trades off some of the positive spherical aberration to produce as large as possible a transfer window in the phase contrast transfer function (see section 5.3.1 for a definition) of non-C_s corrected instruments. Defocus, which as we have seen can be thought of as a wave aberration (it is after all one of the terms in the aberration function expansion, see equation 4.4), is thus used as an adjustable parameters to compensate for C_s, which as Scherzer pointed out was unavoidable. We have now aberration correctors at our disposal: those do not only correct spherical aberration (or C_{50} for the latest generation of instruments) but also most aberrations up to third

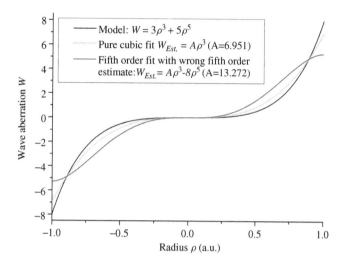

Figure 4.15 Poor man's compensation (a.k.a. 'Super-Scherzer'). A pure cubic fit to a known fifth-order polynomial approximates the original function better than a fifth-order fit with a wrong estimate for the fifth-order coefficient.

order (or fifth order). It therefore allows us to generalise the principle of a Scherzer optimal defocus to finding values for the aberrations that *can* be adjusted to optimally compensate for aberrations that cannot be adjusted (see for example, Chang *et al.*, 2006). Two important principles should be borne in mind. The first is that only aberrations with the same azimuthal dependence can be traded off against one another (i.e. round against round or two-fold against two-fold, etc.). The second is that a bad measurement of the higher-order aberrations is much worse than no measurement at all. If in doubt, set the measured values of the high orders to zero. This is because, if you are estimating a third-order aberration from measurements of an aberration function that includes a large component of fifth-order aberration, then the measured third-order coefficients will be wrong but they will be wrong so that they compensate for the fifth orders almost as well as if you had measured the thirds and fifths correctly and compensated optimally (see Figure 4.15).

4.7 CONCLUSIONS

This has been a brief introduction to the principles and techniques of the current aberration technologies in high resolution STEM, with an attempted emphasis on understanding rather than completeness. To find the literature trail in this modern era of aberration-correction, a personal (and probably inadequate) selection of names to search for are: Haider, M., Hawkes, P., Krivanek, O., Rose, H. and Zach, J. The literature is growing by the day but most publications refer either to one or other author in this list or, at least, to the original papers by Scherzer.

For a more general discussion of electron optics useful books are by, amongst others, Hawkes and Kasper, Grivet, Orloff and Born and Wolf.

REFERENCES

Beck, V.D. (1979) A Hexapole Spherical Aberration Corrector, *Optik*, 53, 241–255.

Born, M. and Wolf, E. (2002) Principles of Optics: Electromagnetic Theory of Propagation, Interference and Diffraction of Light in: *Principles of Optics* (7th edn), Cambridge University Press, Ch. 3 & 4.

Chang L.Y., Kirkland, A.I. and Titchmarsh, J.M. (2006) On the Importance of Fifth-Order Spherical Aberration for a Fully Corrected Electron Microscope, *Ultramicroscopy*, 106, 301–306.

Grivet, P. (1972) *Electron Optics* translated by P.W. Hawkes, revised by A. Septier; 2nd English edn, Pergamon Press, Oxford and New York.

Haider, M. *et al.* (1998) Electron Microscopy Image Enhanced, *Nature*, 392, 768.

Hawkes, P.W., and Kasper, E. (1996) in: *Principles of Electron Optics*, Vol. 2, Academic Press, London and San Diego, Ch. 41.

Hawkes, P.W. (2009) Aberration Correction Past and Present, *Phil. Trans. R. Soc. A*, 367, 3637–3664.

Jackson, J.D. (1975), *Classical Electrodynamics*, John Wiley & Sons Inc, New York.

Kabius, B. *et al.* (2009) First Application of Cc-Corrected Imaging for High-Resolution and Energy-Filtered TEM, *Journal of Electron Microscopy*, 58, 147–155.

Krivanek, O.L., Dellby, N. and Lupini, A.R. (1999) Towards Sub-Angstrom Electron Beams, *Ultramicroscopy*, 78, 1–11.

Leary R. and Brydson R. (2011) Chromatic Aberration Correction: The Next Step in Electron Microscopy in: *Advances in Imaging and Electron Physics*, Vol. 165, pp. 73–130.

Lin, J.A. and Cowley, J.M. (1986) Calibration of the Operating Parameters for an HB5 Stem Instrument, *Ultramicroscopy*, 19, 31.

Orloff, J. (2008) *Handbook of Charged Particle Optics* (2nd edn), CRC Press.

Rose, H. (1981) Correction of Aberrations in Magnetic Systems with Threefold Symmetry, *Nucl. Inst. & Methods*, 187, 187–199.

Ronchi, V. (1964) Forty years of History of a Grating Interferometer, *Appl. Opt.* 3, 437.

Scherzer, O. (1936) Über einige Fehler von Elektronenlinsen, *Zeitschrift Physik*, 101, 593.

Scherzer. O. (1947) Sphärische und chromatische Korrektur von Elektronenlinsen, *Optik*, 2, 114–132.

Williams, D.B. and Carter, C.B. (2009) *Transmission Electron Microscopy* (2nd edn), Springer, New York.

Zach, J. and Haider, M. (1995) Correction of Spherical and Chromatic Aberration In a Low-Voltage SEM, *Optik*, 99, 112.

5

Theory and Simulations of STEM Imaging

Peter D. Nellist

Department of Materials, University of Oxford, Oxford, UK

5.1 INTRODUCTION

As discussed in Chapter 3, the use of an annular dark-field (ADF) detector gave rise to one of the first detection modes used by Crewe and co-workers during the initial development of the modern STEM (Crewe, 1980). The detector consists of an annular sensitive region that detects electrons scattered over an angular range with an inner radius that may be a few tens of mrad up to perhaps 100 mrad, and an outer radius of several hundred mrad (see Chapter 6). The total scattered intensity to this detector is plotted as a function of the probe position during its two dimensional raster scan across the sample thereby forming an image. It has remained by far the most popular STEM imaging mode, and the amount of published data using ADF STEM is increasing rapidly. It was later proposed that, at high enough scattering angles, the coherent effects of elastic scattering could be neglected because the scattering was almost entirely thermal diffuse. This idea led to the use of the high-angle annular dark-field detector (HAADF). In this chapter, we will consider scattering over all angular ranges, and will refer to the technique generally as ADF STEM.

Aberration-Corrected Analytical Transmission Electron Microscopy, First Edition.
Edited by Rik Brydson.
© 2011 John Wiley & Sons, Ltd. Published 2011 by John Wiley & Sons, Ltd.

The purpose of this chapter is to examine the theoretical basis for the image contrast in STEM imaging and in particular ADF and bright-field (BF) imaging. A more extensive discussion of STEM imaging mechanisms has been given by Nellist (2007). Here we will explain what is meant by incoherent imaging, compare it with coherent phase contrast imaging, and explain why the incoherent nature of ADF imaging leads to a powerful imaging mode. We will start by considering scattering from a single atom before going on to consider imaging of a crystal aligned along a high symmetry direction. A major factor in the theory of ADF STEM is the effect of phonon (thermal diffuse) scattering, and we will discuss the strategies for including thermal diffuse scattering (TDS) in the ADF STEM imaging formulation, and the extent to which simulated images have been matched by the experimental data.

Although ADF imaging leads to highly intuitive images (and hence its popularity), the combination of elastic and TDS that gives rise to image contrast makes the theoretical background somewhat complicated. In this chapter, some of the detailed mathematics is presented separately in boxes, with a more intuitive physical explanation presented in the text.

5.2 Z-CONTRAST IMAGING OF SINGLE ATOMS

To develop an understanding of the properties of an annular dark-field image it is appropriate to start by considering the imaging of a single atom. It is immediately apparent from Figure. 5.1, where it can be seen that the hole in the detector is larger than the bright-field (BF) disc, that in the absence of any sample there will be no scattering and therefore no signal detected. The presence of a sample will scatter electrons to the detector and so positive intensity will be seen. We can therefore immediately conclude that this is a dark-field imaging technique where atoms will always show up as bright contrast. Compare this with conventional high-resolution transmission electron microscopy (HRTEM) where both bright and dark contrast can be seen for an atom depending on the exact settings of the imaging optics (see section 9.3).

It should be next noted that the typical collection angles for an ADF detector are relatively large, especially for the so-called high-angle ADF (HAADF) detector. These angles of scatter are much larger than typical Bragg scattering angles, which are typically in the range 10–20 mrad at normal accelerating voltages. In using such a detector it was initially assumed that the intensity of scattering would follow the Rutherford

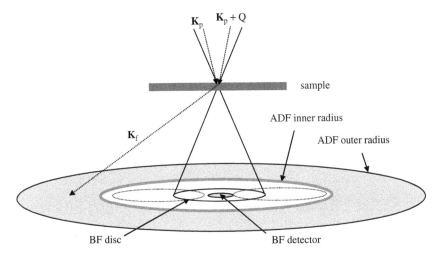

Figure 5.1 A schematic diagram of the interference that results in STEM image contrast. Partial plane-waves in the convergent beam are scattered into a single final wavevector where they interfere.

scattering model of being proportional to Z^2 where Z is the atomic number of the illuminated atom. The Rutherford scattering model assumes an unscreened Coulomb potential, but in practice screening will reduce the Z-dependence a little (see section 1.3). Simulations of scattering using realistic atomic scattering factors suggest that a slightly lower exponent is more reasonable. Nonetheless, this dependence on Z is much stronger than would be observed in HRTEM which uses only the much smaller scattering angles that can be collected by the objective lens. Images taken with HAADF detectors are therefore sometimes referred to as *Z-contrast images*.

If more than one atom is illuminated by the STEM probe, then the detected intensity will obviously increase, leading to images that depend both on atomic number and sample thickness. In practice, however, the degree of coherence can affect how we should add the scattering from multiple illuminated atoms. For example, assuming the Rutherford model that the scattered amplitude from each atom is proportional to Z (so that the scattered intensity is proportional to Z^2), then if the scattering from two atoms, A and B, can be coherently added, we might expect the detected intensity to be proportional to $(Z_A + Z_B)^2$, whereas if we assume incoherent scattering we sum the individual intensities to give $Z_A^2 + Z_B^2$. Despite this complication, HAADF imaging is still widely used to observe changes in atomic composition. By assuming that sudden intensity changes in an image arise from composition changes rather than sharp thickness variations, a qualitative assessment in terms of composition can often be made. An example is shown in Figure 8.5b (Chapter 8) where high intensity features are interpreted in terms of uranium atoms on a carbon support.

Clearly the degree to which the scattering from separate atoms can be treated as being entirely independent of their neighbours (which is the definition of incoherent imaging) is important for the interpretation of ADF imaging, and this will therefore dominate much of this chapter.

5.3 STEM IMAGING OF CRYSTALLINE MATERIALS

We begin now addressing the issue of the degree of coherence in ADF imaging by considering atoms in a crystal lattice and by comparing bright-field (BF) and ADF imaging. Initially we will consider a very thin crystal to avoid having to deal with multiple scattering effects, and we will start by assuming that the lattice is stationary and ignore thermal lattice vibrations.

To calculate the intensity of the scattering that falls on any detector geometry in the STEM for a given illuminating probe position, $\mathbf{R}_p$, we first need to calculate the intensity in the STEM detector plane. This can be calculated by assuming that a sample described by a transmission function, $\phi(\mathbf{R})$ where $\mathbf{R}$ is a two-dimensional position vector in the sample, is illuminated by a STEM probe with complex amplitude, $\psi_p(\mathbf{R}-\mathbf{R}_p)$, located at $\mathbf{R}_p$. The wave exiting the sample will be the product of these two, and so the wave observed at the detector plane (which we can consider as being in the far-field) is given by the Fourier transform of the product of $\phi(\mathbf{R})$ and $\psi_p(\mathbf{R}-\mathbf{R}_p)$. The Fourier transform of this product is given by the convolution between the Fourier transform of $\psi_p(\mathbf{R}-\mathbf{R}_p)$ and the Fourier transform of $\phi(\mathbf{R})$. In equation (4.3) in Chapter 4 we saw that the diffraction limited probe can be written as the inverse Fourier transform of the aperture function, $A(\mathbf{K}_p)$ where $\mathbf{K}_p$ is the transverse component of the wavevector of a partial plane-wave included in the illuminating ensemble (Figure 5.1), and the shift of the probe to position $\mathbf{R}_p$ can be included in reciprocal space by multiplying the aperture function by a linear phase ramp, $\exp(i\mathbf{R}_p \cdot \mathbf{K}_p)$. For a crystal lattice, the Fourier transform of $\phi(\mathbf{R})$ will be a two-dimensional reciprocal lattice of delta-functions with an amplitude given by ϕ_g, which is the gth Fourier component of the specimen transmission function. The convolution between the ϕ_g components and the aperture function then becomes a discrete summation, giving the wave in the detector plane as:

$$\Psi_f(\mathbf{K}_f, \mathbf{R}_p) = \sum_g \phi_g A(\mathbf{K}_f - \mathbf{g}) \exp[-i(\mathbf{K}_f - \mathbf{g}) \cdot \mathbf{R}_p], \qquad (5.1)$$

where $\mathbf{K}_f$ is the transverse component of the wave scattered to the detector plane.

The summation over $\mathbf{g}$ in equation (5.1) describes a series of diffracted discs, which essentially form a convergent beam electron diffraction pattern as one might expect for convergent illumination. If the illuminating objective aperture is sufficiently large, the discs will overlap and interference features will be seen (see Figures 6.14 and 6.15 in Chapter 6 for an example) that are sensitive to the presence of lens aberrations (through the phase of A) and to the probe position (due to the exponential containing $\mathbf{R}_p$). It is the variation of intensity due to these interference effects that give rise to STEM image contrast.

5.3.1 Bright-field Imaging and Phase Contrast

The contrast that may be seen in a bright-field image can be calculated by considering the intensity of the wave in the diffraction plane as a function of $\mathbf{R}_p$ for a small detector located at $\mathbf{K}_f = 0$. It can be shown (see Box 5.3.1a) that such an image may be written:

$$I_{BF}(\mathbf{R}_p) = |\psi_p(\mathbf{R}_p) \otimes \phi(\mathbf{R}_p)|^2 \qquad (5.2)$$

To examine in more detail the transfer of contrast into a BF image, consider a simple case where only the interference between the direct beam and the diffracted beams at $+\mathbf{g}$ and $-\mathbf{g}$ occurs at the detector (Figure 5.2). The complex aperture function, $A(K_p)$, has an modulus that is one inside the aperture and zero outside, and a phase given by the aberration phase shift, $-\chi(K_p)$ (see Appendix 2). Inside the overlap region the moduli of the aperture functions are equal to 1, so equation (5.1) can be written:

$$\Psi_f(\mathbf{K}_f = 0, \mathbf{R}_p) = 1 + \phi_g \exp i[-\chi(-\mathbf{g}) + \mathbf{g} \cdot \mathbf{R}_p]$$
$$+ \phi_{-g} \exp i[-\chi(\mathbf{g}) - \mathbf{g} \cdot \mathbf{R}_p] \qquad (5.3)$$

If we assume that the sample is a weak phase object that shifts the wave passing through the sample by a small phase, σV, the sample

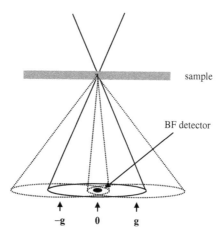

Figure 5.2 A schematic diagram of scattering by a sample that only diffracts into the $+\mathbf{g}$ and $-\mathbf{g}$ beams. For an axial BF detector, the diffracted disc must be sufficiently large that there is overlap between three discs at the detector.

transmission function can be written:

$$\phi(\mathbf{R}) = 1 + i\sigma V(\mathbf{R}) \tag{5.4}$$

Taking the modulus-squared of equation (5.3), and making use of the fact that the projected potential is real (see Box 5.3.1b) results in:

$$I_{\text{BF}}(\mathbf{R}_p) = 1 + 2|\sigma V_{\mathbf{g}}| \cos(\mathbf{g} \cdot \mathbf{R}_p + \angle V_{\mathbf{g}}) \sin \chi(\mathbf{g}) \tag{5.5}$$

This formula corresponds to a set of fringes in the image intensity arising from the phase variation in the object at spatial frequency $\mathbf{g}$. The position of the fringes in the image is controlled by the phase of the $\mathbf{g}^{\text{th}}$ Fourier component of the sample potential, $\angle V_{\mathbf{g}}$. The factor of $\sin \chi(\mathbf{g})$ gives the strength at which each spatial frequency in the phase variation can contribute to the image, and is therefore known as the phase contrast transfer function (PCTF) (see Figure 5.4) which is essentially determined by the complex aperture function. The phase contrast transfer function for BF STEM imaging is identical to that found in high-resolution TEM (see Chapter 9). Phase contrast imaging requires the optimisation of the lens aberrations (usually spherical aberration and defocus) to maximise the PCTF over the widest possible range of spatial frequencies. Such conditions are usually referred to as Scherzer conditions. The analysis presented here verifies the principle of reciprocity as discussed in Chapter 3. A small axial detector in a STEM is equivalent to axial illumination in a CTEM, and so a similar image results.

From Figure 5.2 it immediately becomes apparent that spatial frequencies corresponding to diffraction at angles up to the aperture radius will be transferred. Beyond that spatial frequency there is no disc overlap at the centre of the detector plane, and therefore no contrast will be observed.

Box 5.3.1a

Substituting $K_f = 0$ into equation (5.1) gives:

$$\Psi_f(K_f = 0, R_p) = \sum_g \phi_g A(-g) \exp[ig \cdot R_p] = \mathbf{F}^{-1}[\phi_g A(-g)]$$

where we have identified the summation over g as being the inverse Fourier transform of the product of ϕ_g and A. Taking this inverse Fourier transform leads to a convolution between the inverse Fourier transform of ϕ_g, which by definition is $\phi(R)$, and the inverse Fourier transform of A which is the illuminating probe wavefunction, and the convolution can therefore be written:

$$\Psi_f(K_f = 0, R_p) = \Psi_p(R_p) \otimes \phi(R_p)$$

and taking the intensity as the square of the modulus of Ψ leads to equation (5.2).

Box 5.3.1b

Taking the modulus squared of equation (5.3) and retaining only the term first-order in ϕ (as it is small), gives:

$$|\Psi_f(K_f = 0, R_p)|^2 = I_{BF}(R_p) = 1 + \phi_g \exp i[-\chi(-g) + g \cdot R_p]$$

$$+ \phi_{-g} \exp i[-\chi(g) - g \cdot R_p] + \phi_g^* \exp i[\chi(-g) - g \cdot R_p]$$

$$+ \phi_{-g}^* \exp i[\chi(g) + g \cdot R_p]$$

Because $V(R)$ is real, $V_{-g} = V_g^*$, so using equation (5.4) we have $\phi_g = i\sigma V_g$, $\phi_g^* = -i\sigma V_g^*$, $\phi_{-g} = i\sigma V_g^*$ and $\phi_{-g}^* = -i\sigma V_g$, where V_g is the gth Fourier component of $V(R)$. Substituting these identities, and assuming χ is a symmetric function, gives:

$$|\Psi_f(K_f = 0, R_p)|^2 = I_{BF}(R_p) = 1 + i\sigma V_g \exp i[-\chi(g) + g \cdot R_p]$$

$$+ i\sigma V_g^* \exp i[-\chi(g) - g \cdot R_p] - i\sigma V_g^* \exp i[\chi(g) - g \cdot R_p]$$

$$- i\sigma V_g \exp i[\chi(g) + g \cdot R_p]$$

Collecting terms:

$$I_{BF}(\mathbf{R}_p) = 1 + i(\exp[-i\chi(\mathbf{g})] - \exp[i\chi(\mathbf{g})])$$
$$\times (\sigma V_g \exp[i\mathbf{g} \cdot \mathbf{R}_p] + \sigma V_g^* \exp[-i\mathbf{g} \cdot \mathbf{R}_p])$$
$$= 1 + 2\sigma \sin(\chi(\mathbf{g}))$$
$$\times [|V_g| \exp(i\mathbf{g} \cdot \mathbf{R}_p + \angle V_g) + |V_g| \exp[-i\mathbf{g} \cdot \mathbf{R}_p - \angle V_g]]$$

which simplifies to give equation (5.5).

5.3.2 Annular Dark-field Imaging

In contrast to bright-field imaging, one of the most important character-istics of annular dark-field imaging is its incoherent nature. We start by giving a qualitative explanation of how a large detector can give rise to an incoherent image. Consider elastic scattering of the incident beam by two atoms illuminated by a STEM probe. Because the illuminating probe is highly coherent, and elastic scattering retains this coherence, the inten-sity in the far-field detector plane will show interference features, shown schematically in Figure 5.3 as a set of Young's interference fringes. The actual positions of the interference maxima and minima on the detector will depend on the phase relationship between the scattering from the two atoms. The intensity collected by a small detector, such as a BF detector (Figure 5.3a), will therefore depend on this phase relationship. The sensitivity to phase indicates that BF is a coherent imaging process and qualitatively explains how phase contrast imaging can be achieved.

A large detector, such as an ADF detector (Figure 5.3b), will average over many of the interference features in the detector plane suppressing any coherent phase information. The intensity collected will simply depend on the intensity of the STEM probe that illuminates each atom, and the fraction of this intensity that is scattered to the detector. The image can therefore be written as a convolution of the intensity of the STEM probe and an object function, $O(\mathbf{R})$ that represents the fraction of intensity that each atom is able to scatter to the detector (which will be a strong function of Z for a HAADF detector), hence:

$$I_{ADF}(\mathbf{R}_p) = |\psi_p(\mathbf{R}_p)|^2 \otimes O(\mathbf{R}_p) \tag{5.6}$$

For a more quantitative analysis that allows $O(\mathbf{R}_p)$ to be defined, and ignoring for the moment any effects of TDS, the ADF image intensity

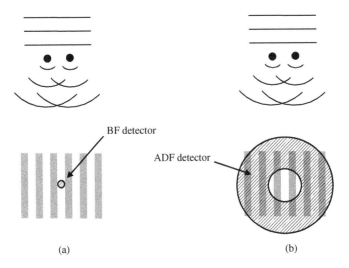

BF detector

ADF detector

(a)   (b)

Figure 5.3 The interference between the scattering from two coherently illuminated atoms will lead to interference features in the detector plane (far-field) as shown schematically here. A small detector will be sensitive to the position of the fringes, and therefore to the phases of the scattered waves. A larger detector sums over several fringes thereby removing any phase sensitivity.

is found by simply integrating the detector plane intensity (the square of the wave or wavefunction amplitude) over the extent of the detector area. We can define a detector function, $D_{ADF}(K_f)$, as having a value of unity for all values of K_f collected by the ADF detector, and zero elsewhere. Starting from equation (5.1), the ADF intensity as a function of probe position can then be written:

$$I_{ADF}(R_p) = \int D_{ADF}(K_f) \left| \sum_g \phi_g A(K_f - g) \exp[-i(K_f - g) \cdot R_p] \right|^2 dK_f$$

$$(5.7)$$

It is easier now to consider the Fourier transform of the image. Rather than writing the image intensity as a function of the probe position, R_p, we take the Fourier transform so that the intensity is written as a function of spatial frequency in the image, Q. Following the derivation in Box 5.3.2a, we find that the intensity as a function of Q can be written:

$$I_{ADF}(Q) = \int D_{ADF}(K_f) \sum_g \phi_g \phi^*_{g-Q} A(K_f - g) A^*(K_f - g + Q) \, dK_f \quad (5.8)$$

An interpretation of equation (5.8) is that for a spatial frequency, Q, to show up in the image, two beams incident on the sample separated by Q must be scattered by the sample so that they end up in the same final wavevector K_f where they can interfere (Figure. 5.1). One such beam is scattered by the sample through a reciprocal lattice vector, g, and the other by $g-Q$, thus arriving in the same final wavevector, K_f. This model of STEM imaging is applicable to any imaging mode, even when TDS or inelastic scattering is included. Both these incident beams must lie within the objective aperture, and so we can immediately conclude that STEM is unable to resolve any spacing smaller than that allowed by the diameter of the objective aperture, no matter which imaging mode is used. Note that this upper bound to the resolution limit is twice that for BF imaging, which can be explained by realising that this imaging mode only requires the interference of two beams in the incident cone, rather than the three required by BF imaging.

While both the scattering vector g and the wavevector at the detector, K_f, may be large, the quantities $K_f - g$ and $K_f - g + Q$ must lie within the aperture radius, so must be small relative to the size of the detector. It therefore follows that we can expect that the aperture overlap region to be small compared with the physical size of the ADF detector. Making this approximation, equation (5.8) can be rewritten in the form of equation (5.6) (see Box 5.3.2b).

Let us now summarise the important differences between coherent and incoherent imaging. For HRTEM and BF STEM imaging, a specimen transmission function is convolved with the complex wavefunction of the illuminating probe (equation 5.2), and the intensity measured is the modulus-squared of this convolution. In this imaging model, the importance of the phase of the illuminating probe and the sample transmission function show that interference processes are important, and this is known as the *coherent imaging* model. Simply changing the phase variations within the probe function, $\psi_p(R)$, by changing the defocus setting for example, can reverse the contrast in a coherent image. Although coherent images contain a wealth of information, their interpretation requires careful determination of the imaging parameters and subsequent image simulations to match to the experimental data.

Conversely, equation (5.6) is known as the *incoherent imaging* model. Because only intensities are involved, it shows that interference between scattering from spatially separated parts of the sample does not influence the image contrast. The strength of incoherent imaging is that is leads to images which are relatively straightforward to interpret. The contrast reversal and delocalisation effects seen in coherent TEM imaging cannot occur, and bright contrast in ADF images can in general be interpreted as

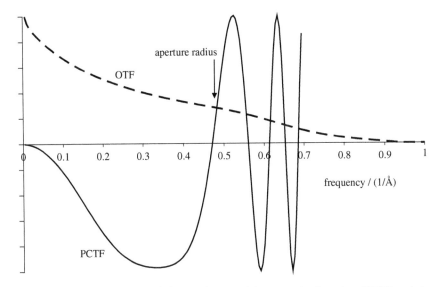

Figure 5.4 A comparison of the incoherent object transfer function (OTF) and the coherent phase-contrast transfer function (PCTF) for identical imaging conditions ($V = 300$ kV, $C_S = 1$ mm, $z = -40$ nm).

the presence of atoms or atomic columns. By reciprocity, use of the ADF detector can be compared to illuminating the sample with large angle incoherent illumination (see also Chapter 6). In optics, the Van Cittert-Zernicke theorem (Born and Wolf, 2002) describes how an extended source gives rise to a coherent envelope that is the Fourier transform of the source intensity function. An equivalent coherence envelope exists for ADF imaging, and is the Fourier transform of the detector function, $D(K_f)$. As long as this coherence envelope is significantly smaller than the probe function, the image can be written in the form of equation (5.6) as being incoherent. This condition is the real-space equivalent of the approximation used in Box 5.3.2b.

The strength at which a particular spatial frequency in the object is transferred to the image is known, for incoherent imaging, as the optical transfer function (OTF). The OTF for incoherent imaging, $T(Q)$, is simply the Fourier transform of the probe intensity function. In general it is a positive, monatonically decaying function (see (Black and Linfoot, 1957) for examples under various conditions), which compares favourably with the phase contrast transfer function, PCTF, for the same lens parameters (Figure. 5.4). In particular, the incoherent OTF can be seen to extend to twice the spatial frequency of that corresponding to the aperture radius, which means that a spatial resolution corresponding to the diameter of the objective aperture can be achieved.

Box 5.3.2a

Expanding the modulus-squared in Equation (5.7) gives a double summation over Fourier components of ϕ:

$$I_{ADF}(\mathbf{R_p}) = \int D_{ADF}(\mathbf{K_f}) \sum_{g,h} \phi_g \phi_h^* A(\mathbf{K_f} - \mathbf{g}) A^*(\mathbf{K_f} - \mathbf{h}) \exp[i(\mathbf{g} - \mathbf{h}) \cdot \mathbf{R_p}] \, d\mathbf{K_f}.$$

A simplification can now be made by taking the Fourier transform of the ADF image with respect to $\mathbf{R_p}$ to give a function in terms of image spatial frequency, $\mathbf{Q}$:

$$I_{ADF}(\mathbf{Q}) = \int D_{ADF}(\mathbf{K_f}) \sum_{g,h} \phi_g \phi_h^* A(\mathbf{K_f} - \mathbf{g}) A^*(\mathbf{K_f} - \mathbf{h})$$

$$\times \int \exp[i(\mathbf{g} - \mathbf{h} - \mathbf{Q}) \cdot \mathbf{R_p}] \, d\mathbf{R_p} \, d\mathbf{K_f}.$$

Performing the $\mathbf{R_p}$ integral results in a delta function:

$$I_{ADF}(\mathbf{Q}) = \int D_{ADF}(\mathbf{K_f}) \sum_{g,h} \phi_g \phi_h^* A(\mathbf{K_f} - \mathbf{g}) A^*(\mathbf{K_f} - \mathbf{h}) \delta_{g-h-Q} \, d\mathbf{K_f},$$

allowing the $\mathbf{h}$ summation to be dropped because the terms in the summation are only non-zero when $\mathbf{h} = \mathbf{g} - \mathbf{Q}$.

Box 5.3.2b

Substitute $\mathbf{K_f} - \mathbf{g}$ for an incident wavevector, $\mathbf{K_p}$ in equation (5.8):

$$I_{ADF}(\mathbf{Q}) = \int D_{ADF}(\mathbf{K_p} + \mathbf{g}) \sum_{g} \phi_g \phi_{g-Q}^* A(\mathbf{K_p}) A^*(\mathbf{K_p} + \mathbf{Q}) \, d\mathbf{K_p}.$$

If we assume that the geometry of the ADF detector (i.e. the inner radius and the difference between the inner and outer radii) is much larger than the extent of $\mathbf{K_p}$ allowed by the product of the aperture functions, then we can ignore the $\mathbf{K_p}$ dependence of D_{ADF}, thus allowing separation of the integral and summation:

$$I_{ADF}(\mathbf{Q}) = \sum_{g} D_{ADF}(\mathbf{g}) \phi_g \phi_{g-Q}^* \int A(\mathbf{K_p}) A^*(\mathbf{K_p} + \mathbf{Q}) \, d\mathbf{K_p}.$$

In making this approximation we have assumed that the contribution of any overlap regions that are partially detected by the ADF detector is small compared with the total signal detected. The integral containing the aperture functions is actually the autocorrelation of the aperture function. The Fourier transform of the probe intensity is the autocorrelation of A, thus taking the inverse Fourier transform results in a convolution with the probe intensity, as given in equation (5.6). The remainder of the expression can be identified as the Fourier transform of the object function that is convolved with the probe intensity in equation (5.6). The effect of the detector is to provide a high-pass spatial filter on the specimen transmission function, $\phi(\mathbf{R})$. The object function, $O(\mathbf{R_p})$, is then arrived at by taking the modulus-squared of this filtered transmission function.

5.4 INCOHERENT IMAGING WITH DYNAMICAL SCATTERING

The derivation in section 5.3 assumed that the sample could be represented by a multiplicative transmission function, and the scattering by such a sample is not dependent on small variations in the incident beam direction. For many, if not most, of the samples examined using (S)TEM this is not the case. Electrons interact strongly with matter, and are multiply elastically scattered. This multiple scattering, combined with propagation of the beams in the sample between scattering events, leads to dynamical scattering (see Chapter 1). Dynamical scattering is sensitive to beam tilt, and an important question is whether our derivation of incoherence remains valid under dynamical scattering conditions.

Given that the integration over a large detector can destroy the coherence of the imaging process even when we have only included coherently scattered Bragg electrons, we might also expect that the coherence is similarly destroyed when we include the effects of coherent dynamical scattering. Consider once again that the form of the intensity at the detector plane is a coherent convergent electron diffraction (CBED) pattern. It is well known that the effect of dynamical diffraction on the CBED pattern is to introduce structure in each of the diffracted discs (Spence and Zuo, 1992), showing that the diffraction condition is changing as a function of incident beam angle within the incident cone of illumination. The summation over the large ADF detector will once again average over the structure arising from the dynamical scattering, suggesting that the complicating effects of dynamical scattering will be reduced. Indeed, this explains one of the powerful features of ADF incoherent imaging, that directly interpretable images are formed even in relatively thick samples where strong dynamical diffraction would complicate the interpretation of any coherent imaging technique such as phase contrast HREM.

To examine this effect more quantitatively, a Bloch wave analysis has been performed (Nellist and Pennycook, 1999) to calculate the intensity when a large STEM detector is used. Bloch waves are the eigenfunctions found when solving the Schrödinger equation for the fast electron in a crystal. The image contrast is found to arise from interference terms between pairs of Bloch waves, but that the contribution of these terms to the ADF contrast is suppressed when the Bloch waves have significant wave amplitude that is separated by more than the coherence envelope imposed by the detector. Although the neglect of TDS in this calculation means that image intensities are not accurately modelled, it does confirm that interference effects from parts of the sample laterally separated are not observed, similar to that seen for the thin sample in the previous section. The Bloch wave analysis also shows that the states that contribute most strongly to the intensity collected by the ADF detector are the states most localised on the atomic columns (and, by analogy with atoms, labelled as 1s states), and these are known as *channelling* states.

Electron channelling, which was introduced in Chapter 1, plays an important role in STEM imaging. A schematic diagram of the channelling process is shown in Figure. 5.5. When the illuminating probe is located over an atomic column in a crystal, the electrons are to some degree trapped by the attractive potential of the atoms. The process is somewhat like a waveguide, and oscillations of the intensity of the wave on the atom sites are seen as a function of depth in the crystal. From a Bloch wave point of view, the channelling arises from the strong excitation of

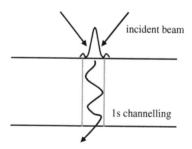

Figure 5.5   When the incident probe illuminates an atomic column, the electrons are attracted by the electrostatic potential and channel along the column. The effect is somewhat similar to a waveguide, and the intensity of the wave at the centre of the column oscillates in strength. Due to strong absorption of the channelling wave, only a small number of such oscillations are seen before the wave is completely depleted.

the most localised 1s-type states (Pennycook and Jesson, 1990) and the depth oscillations arise from the interference between these states and the ensemble of other states that have significant amplitude at the atom sites. In the previous paragraph we discussed how these states contribute most strongly to the ADF signal, and an ADF image can be thought of as a map of the strength of channelling as a function of illuminating probe position. When a probe is located over a column, strong channelling occurs and a bright intensity is seen.

Channelling can be disrupted by defects such as dislocations, or indeed simply strain in the lattice, and such defects can also cause contrast in ADF images. The exact influence of lattice defects or strain on an image is not intuitively predictable. For example, at low detector inner radii strain can cause bright contrast, whereas for HAADF strain generally leads to a darkening of the image (Yu *et al.*, 2004).

5.5 THERMAL DIFFUSE SCATTERING

So far we have only considered elastic scattering to the ADF detector, and in the case of crystalline samples this will be in the form of Bragg diffracted beams. At the angles collected by a typical ADF detector, an effect of the thermal lattice vibrations is to attenuate the strength of the Bragg beams with the intensity instead appearing as a diffuse intensity known as thermal diffuse scattering (TDS). The attenuation of the Bragg beams is usually included through the use of a Debye-Waller factor. Scattering by thermal phonons is an inelastic scattering process because it involves the generation or destruction of phonons. The thermal energy losses involved (of the order of tens of meV) are much smaller than can be detected using modern energy-loss spectrometers, and the scattering is often referred to as being quasi-elastic because it cannot be distinguished from the elastic scattering. It has been shown (Amali and Rez, 1997) that the TDS intensity integrated over an HAADF detector exceeds that of the elastic scattering (including HOLZ reflections), and so any theoretical or computational treatment of ADF imaging must include TDS if it is to be complete.

Because TDS involves inelastic scattering processes, it is often regarded as being an incoherent scattering process. By this we mean that the scattering from separate atoms cannot interfere coherently. The domination of TDS in the ADF signal has been used as a justification (Howie, 1979) for the incoherent nature of ADF imaging. In section 5.3.2, however, we asserted that the incoherence arises from the extended geometry of the detector. These two views of the origins of incoherence have caused the literature on the theory of ADF to be somewhat contradictory and confusing. To reconcile these points of view is beyond the scope of this chapter, but here we note that the detector geometry always imposes incoherence no matter what the nature of the scattering.

5.5.1 Approximations for Phonon Scattering

Early treatments of the effects phonon scattering were within the context of electron diffraction and diffraction contrast imaging (Hall and Hirsch, 1965) where it was assumed that once the primary electrons had been phonon scattered, that they no longer played any role in the diffraction or imaging, and were treated as being 'lost'. This approach led to the idea of absorption, and even today we refer to multiple scattering calculations as 'including absorption'. In fact, very few electrons are actually stopped by the sample, and therefore 'absorption' is somewhat of a misnomer. In particular for ADF imaging, because of the large angles collected by the detector, a reasonable fraction of the phonon-scattered electrons are incident on the detector, and so the 'absorption' is actually what we are detecting.

Calculation of the 'absorption' is easiest when the atoms in a crystal are assumed to be thermally vibrating entirely independently of their neighbours. In terms of phonon modes of vibration, this approximation, referred to as the *Einstein dispersion approximation* assumes that all phonon modes have the same energy. Amali and Rez (1997) have reproduced the initial (Hall and Hirsch, 1965) approach in the context of ADF imaging. By calculating the diffraction pattern for the crystal, and then taking a time average of the intensity over the random vibrations, the attenuation of the elastically scattered beam can be found. Subtracting the attenuated diffraction pattern from that observed for a stationary lattice, and assuming that the apparently 'absorbed' electrons at a given angle will end up being at the same angle after phonon scattering, the phonon scattered intensity may be calculated in terms of a scattering cross-section (Pennycook, 1989; Mitsuishi *et al.*, 2001).

In practice, longer wavelength phonon modes in the crystal will have lower energies and will therefore be more strongly excited. Having longer wavelength phonons present means that the thermal vibrations of neighbouring atoms are likely to have some degree of correlation, suggesting a degree of coherence in the phonon scattering from the lattice. Jesson and Pennycook (1995) performed a theoretical study of the effect of realistic phonon distributions and showed that a coherence length could be derived that was typically a few atomic spacings in length.

The suggestion made by Jesson and Pennycook (1995) that a realistic phonon dispersion does not render the thermal scattering from neighbouring atoms as being completely incoherent is highly relevant to the present discussion given that the initial justification for the use of an HAADF detector is to collect only thermal scattering which was assumed to be incoherent. We have already seen in section 5.3.2 that the coherence in the transverse direction (i.e. perpendicular to the beam optic axis) is very effectively destroyed by the large dimensions of the detector. This point was similarly made by Muller *et al.* (2001), but it is less clear to what degree both the detector geometry and the TDS can destroy coherence along the optic axis. In a case of perfect incoherence, we might expect the HAADF signal to increase linearly in thickness. In both simulations and experimental data (Hillyard and Silcox (1993), Figure. 5.6) although a monatonic increase in intensity with respect to thickness is seen, it is far from being linear, though this deviation is most likely attributable to the channelling rather than any effect of partial coherence. For example, if we assume that the scattering to the HAADF detector is perfectly incoherent, then the contribution of each atom in a thickness of sample will depend on the illuminating wavefunction intensity at that atom. The total intensity will therefore be a summation through the thickness, or equivalently, we can say that the rate of increase in intensity as a function of depth depends on the wavefunction intensity at that depth. From the channelling oscillations, we can see that we have depths where the intensity on the column is high, and others where is it lower. Thus we might expect to have periodically varying

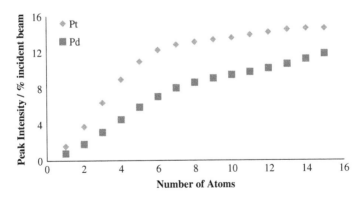

Figure 5.6 A calculation of the peak ADF image intensity (expressed as a fraction of the incident beam current) for isolated Pt and Pd columns expressed as a function of number of atoms in a column. (Microscope parameters: 200 kV; aberration-free; 24 mrad semi-angle of convergence; ADF detector 90–200 mrad). (Courtesy of H.E. (private communication).)

slopes of a plot of column intensity versus thickness, which are usually observed in simulations. The s-state will be very strongly absorbed due to its localization on the column. Typically only two or three channelling oscillations are seen before the state is entirely absorbed (which means scattered by phonons). The intensity versus thickness plot therefore has an overall reduction in slope, until it saturates when the s-state is depleted (Figure 5.6).

It must be emphasised that the description above has many approximations and only gives us a qualitative description of HAADF intensities. For a more quantitative description we must perform detailed calculations including the effect of absorption.

5.6 METHODS OF SIMULATION FOR ADF IMAGING

The use of simulations in ADF STEM is still not nearly as widespread as it is in HRTEM (see Chapter 9). The main reason for this is that the incoherent nature of ADF STEM imaging means that often the images can be directly interpreted in terms of the projected structure, with higher Z atomic columns showing up brighter. Furthermore, the complexities associated with needing to include TDS scattering have acted as an impediment to quantitative interpretation. Nonetheless, ADF STEM simulations are now starting to be used more extensively.

From the discussion in sections 5.3 and 5.3, it is clear that any simulation method must include the effects of dynamical scattering of the probe and also the contribution of thermal diffuse scattering. Because atomic resolution ADF STEM is very often performed on crystals aligned along high symmetry directions where the atoms form resolvable columns, the effects of channelling can be very significant. The approaches to including both channelling and TDS in the calculations fall into two main categories: those that use absorptive potentials and those that make use of the frozen phonon approach.

5.6.1 Absorptive Potentials

The absorptive potential approach is discussed in detail by Allen *et al.* (2003) and Findlay *et al.* (2003), and makes the following assumptions:

1. That the scattering to the ADF detector is dominated by TDS.
2. That the TDS from each atom can be assumed to be incoherent with respect to neighbouring atoms. In the direction transverse to the beam,

the size of the ADF detector ensures that this is a good approximation, as discussed in section 5.3.2. Along the beam direction we rely more on the predominant detection of TDS to break the coherence and it is possible that some degree of coherence between atoms may remain.

3. That once thermally scattered, the electrons undergo no further elastic scattering by the lattice. It is therefore assumed that Kikuchi lines do not affect the intensity arriving at the detector.

Under these incoherent approximations, the total TDS intensity at the ADF detector can then be written in terms of a local scattering function, $\phi'_{ADF}(\mathbf{r})$, due to the sample, multiplied by the intensity of the electron wavefunction in the crystal:

$$I_{ADF}(\mathbf{R_p}) = \int \phi'_{ADF}(\mathbf{r})|\psi(\mathbf{r}, \mathbf{R_p})|^2 \, d\mathbf{r}. \qquad (5.9)$$

where $\psi(\mathbf{r}, \mathbf{R_p})$ is the wavefunction in the crystal for an illuminating probe at lateral position $\mathbf{R_p}$, and the integral is in three-dimensional position, $\mathbf{r}$, over the whole of the crystal.

In order to evaluate equation (5.9) the wavefunction in the crystal is required and either a Bloch wave approach (Nellist and Pennycook, 1999) or a multislice calculation (Kirkland, 1998) can be used (see Chapter 1). For both the Bloch wave approach and the multislice approach, the elastic wave will be attenuated by the TDS and other incoherent scattering processes. Because this scattering can be treated as being incoherent, it can also be treated using a local potential, $\phi'_{tot}(\mathbf{r})$. It is important to note that $\phi'_{tot}(\mathbf{r})$. includes all processes that reduce the intensity of the elastic wave, including all of the TDS, and is generally referred to as the absorption potential. In the case of ADF imaging, many of these TDS scattered electrons are detected, hence the need to also compute $\phi'_{ADF}(\mathbf{r})$, which is the part of $\phi'_{tot}(\mathbf{r})$ that makes it to the detector.

It is also possible to include the small amount of elastic scattering to the ADF detector by using the Fourier transform of the exit-surface wave, which may be important if low angle ADF detectors are being considered.

5.6.2 Frozen Phonon Approach

The frozen phonon approach (Kirkland, 1998) is based on the realization that an electron accelerated to a typical energy of 100 keV is traveling at about half the speed of light. It therefore transits a sample of thickness, say, 10 nm in 3×10^{-17} s, which is much smaller than the typical period of a lattice vibration ($\sim 10^{-13}$ s). Each electron that transits the sample will see a lattice in which the thermal vibrations are frozen in some configuration, with each electron seeing a different configuration. Multiple multislice calculations can be performed for different thermal displacements of the atoms, and the resultant intensity in the detector plane summed over the different configurations (see also Koch (2010)).

The frozen phonon multislice method is not limited to calculations for STEM; it can be used for many different electron scattering experiments. In STEM, it will give the intensity at any point in the detector plane for a given illuminating probe position. The calculations faithfully reproduce the TDS, Kikuchi lines and higher-order Laue zone (HOLZ) reflections. To compute the ADF image, the intensity in the detector plane must be summed over the detector geometry, and this calculation repeated for all the probe positions in the image. The frozen phonon method can be argued to be the most complete method for the computation of ADF images. Its major disadvantage is that it is computationally expensive. For absorptive potential simulations of ADF STEM, one calculation is performed for each probe position. In a frozen phonon calculation, several (typically in the range 10–100) multislice calculations are required for each probe position in order to average effectively over the thermal lattice displacements.

Recently there has been a greater effort in quantitatively interpreting ADF STEM by comparison with simulations, and this has led to careful testing of the predictions of simulations against experiments. It has been shown that the absorption potential approach is effective, but only at relatively low thicknesses where multiple phonon or phonon-elastic scattering can be neglected (LeBeau et al., 2008). At higher thicknesses, a frozen-phonon calculation is required, but the fit to the experimental data is good.

5.7 CONCLUSIONS

In summary, the incoherent nature and Z-contrast observed in ADF STEM make it an extremely powerful technique for the atomic resolution characterisation of materials, and its popularity has grown significantly over the last fifteen years or so. The main reason for this growth of interest is the relatively simple qualitative interpretation of the images. Care must be taken, however, because the effects of channelling can lead to effects that may be misinterpreted, for example strain contrast being confused with a composition variation.

Given the generally readily interpretable nature of the images, it is somewhat disappointing that simulations are not more widely used to provide quantitative information. This probably mostly attributable to direct interpretability of the images that often render simulations unnecessary, but the necessity of including TDS in the calculations certainly adds to the complexities compared to HRTEM simulation methods.

REFERENCES

Allen, L.J., Findley, S.D., Oxley, M.P. and Rossouw, C.J. (2003) Lattice-Resolution Contrast From a Focused Coherent Electon Probe. Part I. *Ultramicroscopy*, 96, 47–63.

Amali, A. and Rez, P. (1997) Theory of Lattice Resolution in High-angle Annular Dark-field Images. *Microscopy and Microanalysis*, 3, 28–46.

Black, G. and Linfoot, E.H. (1957) Spherical Aberration and the Information Limit of Optical Images. *Proceedings of the Royal Society (London) Ser. A*, 239, 522–540.

Born, M. and Wolf, E. (2002) *Principles of Optics* (7th edn), Cambridge University Press.

Crewe, A.V. (1980) The Physics of the High-Resolution STEM. *Reports of Progress in Physics*, 43, 621–639.

Hall, C.R. and Hirsch, P.B. (1965) Effect of Thermal Diffuse Scattering on Propagation of High Energy Electrons Through Crystals. *Proceedings of the Royal Society (London) Ser. A*, 286, 158–177.

Hillyard, S. and Silcox, J. (1993) Thickness Effects in ADF STEM Zone Axis Images. *Ultramicroscopy*, 52, 325–334.

Howie, A. (1979) Image Contrast and Localised Signal Selection Techniques. *Journal of Microscopy*, 117, 11–23.

Jesson, D.E. and Pennycook, S.J. (1995) Incoherent Imaging of Crystals Using Thermally Scattered Electrons. *Proceedings of the Royal Society (London) Ser. A*, 449, 273–293.

Kirkland, E.J. (1998) *Advanced Computing in Electron Microscopy*, New York, Plenum.

Koch, C.T. (2010) QSTEM [Online]. Available: http://www.christophtkoch.com/stem/index.html [Accessed April 2011].

LeBeau, J.M., Findlay, S.D., Allen, L.J. and Stemmer, S. (2008) Quantitative Atomic Resolution Scanning Transmission Electron Microscopy. *Physical Review Letters*, 100, 4.

Mitsuishi, K., Takeguchi, M., Yasuda, H. and Furuya, K. (2001) New Scheme For Calculation of Annular Dark-Field STEM Image Including Both Elastically Diffracted and TDS Wave. *Journal of Electron Microscopy*, 50, 157–162.

Muller, D.A., Edwards, B., Kirkland, E.J. and Silcox, J. (2001) Simulation of Thermal Diffuse Scattering Including a Detailed Phonon Dispersion Curve. *Ultramicroscopy*, 86, 371–380.

Nellist, P.D. (2007) Scanning Transmission Electron Microscopy. *In*: Hawkes, PW. and Spence, J.C.H. (eds) *Science of Microscopy*. Springer.

Nellist, P.D. and Pennycook, S.J. (1999) Incoherent Imaging Using Dynamically Scattered Coherent Electrons. *Ultramicroscopy*, 78, 111–124.

Pennycook, S.J. (1989) Z-Contrast STEM for Materials Science. *Ultramicroscopy*, 30, 58–69.

Pennycook, S.J. and Jesson, D.E. (1990) High-Resolution Incoherent Imaging of Crystals. *Physical Review Letters*, 64, 938–941.

Spence, J.C.H. and Zuo, J.M. (1992) *Electron Microdiffraction*, New York, Plenum Press.

Yu, Z., Muller, D.A. and Silcox, J. (2004) Study of Strain Fields at a-Si/c-Si Interface. *Journal of Applied Physics*, 95, 3362–3371.

6

Details of STEM

Alan Craven

Department of Physics and Astronomy, University of Glasgow, Glasgow, Scotland, UK

In Chapter 4, the method for obtaining an objective lens corrected for geometric aberrations up to 5th order was discussed. In Chapter 5 the theory behind high spatial resolution imaging was considered. In this chapter, the other elements of the STEM optical column needed to make best use of this corrected objective lens are considered in greater detail. Section 6.1 considers the current required to get statistically significant data while section 6.2 considers the relationship between probe size, probe current and probe angle. The key components in the column are then considered starting at the gun. Section 6.3 considers the features needed in the gun and condenser system to allow the corrected objective lens to form the desired probe. Section 6.4 considers the scanning system required to control the probe position while section 6.5 covers the requirements of the specimen stage needed to place the specimen in the objective lens at the correct position and in the correct orientation. After the electrons have passed through the specimen, the scattering distribution must be transferred to the detectors using post-specimen optics and these are considered in section 6.6. After a short discussion of beam blanking in section 6.7, the detectors themselves are considered in section 6.8. Section 6.9 considers the form of the scattered electron distribution, the implications of having a coherent probe and how detectors can be deployed to extract information from the scattering distribution. Section 6.10 considers the signal acquisition itself and how to obtain the maximum information possible from each incident electron.

Aberration-Corrected Analytical Transmission Electron Microscopy, First Edition.
Edited by Rik Brydson.
© 2011 John Wiley & Sons, Ltd. Published 2011 by John Wiley & Sons, Ltd.

6.1 SIGNAL TO NOISE RATIO AND SOME OF ITS IMPLICATIONS

As introduced in Chapter 1, at the end of section 1.3, for both imaging and analysis, the limiting factor is the signal to noise ratio (SNR) available compared to the contrast in an image or the characteristic signal to background ratio in an analytical spectrum. For example, if the signal in one pixel of an image is S_F and that in an adjacent pixel is $S_F + \Delta S$, then ΔS must be bigger than N, the RMS noise in the signal, by some factor if the two pixels are to be perceived as having a different intensity. In the early days of television, the Rose criterion of $\Delta S > 5N$ was adopted (Rose, 1948). Ideally, detectors should not add noise. In this ideal case, if n signal electrons contribute to the pixel, the noise is simply the Poisson noise, i.e. $\sqrt{n}$. Thus, if the contrast, $\Delta S/S_F$, available in a mode of operation is known, the number of signal electrons required in each pixel is also known, i.e. $n > 25(S_F/\Delta S)^2$. The mode of operation also determines the number of signal electrons per incident electron and hence the number of incident electrons required per pixel can be calculated. Such calculations need to be done on a case by case basis. For instance, in a bright field image, the number of signal electrons is a significant fraction of the number of incident electrons but the inherent contrast is low while, in a high angle annular dark field (HAADF) image, the number of signal electrons is a small fraction of the number of incident electrons but the inherent contrast is high.

If the microscope system were stable enough and we had the patience, the required number of signal electrons could be acquired by collecting for a sufficiently long time whatever the incident electron current. In reality, neither condition is fulfilled. The measurements need to be made in an acceptable time. My patience wears thin after about twenty minutes of acquisition but it can be stretched if it is important. During acquisition, the system can change for a number of reasons, e.g. the specimen can drift and currents and voltages change due to temperature fluctuations in the air and cooling water. This causes relative motion between the probe and the specimen. Thus there is a characteristic time within which the signal must be recorded and the smaller the pixel, the shorter the time available. Thus the current available in the probe is an important factor. The current required affects the size of probe required in practice and hence the spatial resolution.

Even if the required current is available, the sample can be unstable in the electron beam, e.g. it can change its structure, be sputtered away or grow a contamination layer. As mentioned in section 1.5, such electron beam induced damage depends primarily on the fluence (i.e. number of electrons per unit area) passing through the specimen, but can also depend on flux. Additionally, where a temperature rise is involved in the damage process, the total incident current can be the key factor.

In many cases, there will be a maximum fluence that can be given to the specimen whilst retaining it sufficiently close to its original state that the results are valid. Since the number of incident electrons required in each pixel is determined by the contrast, the limit to the number of electrons per unit area will define the minimum area to be irradiated and, hence, this may itself limit the ultimate spatial resolution.

6.2 THE RELATIONSHIPS BETWEEN PROBE SIZE, PROBE CURRENT AND PROBE ANGLE

6.2.1 The Geometric Model Revisited

Chapter 2, section 2.6 discusses the relationship between probe diameter, d, probe current, I_{probe}, and probe semi-angle, α in terms of the geometric model (i.e. the square root of the sum of the squares of the contributions from the individual effects). A wave optical treatment is required to deal with this properly, as discussed in Chapter 4. However, it is instructive to re-visit the geometric approximation after the discussion of aberration correction in Chapter 4. The diameters of the contributions are typically a Gaussian image of the source, d_s, a diffraction disc, d_d, and a geometric aberration disc, d_g.

The contributions from the aberrations are considered first and, for simplicity, only the aberrations with cylindrical symmetry are considered, i.e. in the notation of Chapter 4, only those of the form C_{NS} with $N = n$ and $S = 0$ (i.e. $C_{n,0}$) with n odd The aberration with n equal 3 is spherical aberration. This causes the ray inclined at $\theta = \alpha$ to the axis to cross the Gaussian image plane a distance $C_{3,0}\alpha^3$ from the Gaussian image point (here taken to be on the axis) and crosses the axis a distance $C_{3,0}\alpha^2$ before the Gaussian image plane,[1] i.e. the spherical aberration always causes overfocus in a round electron lens. Here $C_{3,0}$ is used rather than C_S to allow generalisation to n^{th} order, as in Chapter 4. Thus one might think that d_g appropriate to spherical aberration would be $2C_{3,0}\alpha^3$. However, the bundle of rays has a narrower waist, known as a disc of least confusion, of diameter $1/2 C_{3,0}\alpha^3$ at a distance $3/4 C_{3,0}\alpha^2$ from the Gaussian image plane. Thus some defocus is used to minimise

---

[1] Note in Figure 4.2 the overfocus of aberrated rays is equal to $(C_{3,0}\theta^3)/\tan\theta = C_{3,0}\theta^2$ for small θ.

the effect of the spherical aberration. It is easy to extend this to an nth order aberration giving d_g of the form $A_n C_{n,0} \alpha^n$ where $C_{n,0}$ is the aberration coefficient and A_n is a numerical constant that gives the diameter of the disc of least confusion. For the 3rd order case A_3 is just $1/2$ and it increases towards unity as the order of the aberration increases. Chromatic aberration can be included if n is taken as 1 and A_1 as $\Delta E / E_0$.

The various contributions to the probe size as a function of α are shown in Figure 6.1 and demonstrate that one aberration is normally dominant in a particular region of α. In Figure 6.1, the values of C_C, $\Delta E / E_0$ (3×10^{-6}) and $C_{3,0}$ used to construct the plot are typical of the uncorrected aberrations of SuperSTEM1 at Daresbury Laboratories. The values of $C_{5,0}$ and $C_{7,0}$ are estimates of the uncorrected 5th and 7th order aberrations respectively. Before correction, the absolute minimum probe size occurs where the lines for diffraction and $C_{3,0}$ cross. After correction of $C_{3,0}$, the absolute minimum probe size occurs where the lines for diffraction and C_C cross. It can be seen that, using this geometric model, correction of $C_{5,0}$ will not lead to a reduction of the minimum probe size because it not the limiting aberration. However, strictly this needs to be investigated properly using the wave optical treatment in order to determine the situation in practice (e.g. Krivanek et al., 2008a).

The size of the Gaussian image is controlled by the demagnification between the source and the probe, and can be expressed in terms of the source brightness, B, as given by equation (2.10). The diffraction term is taken as the diameter of the Airy disc. This results in the following

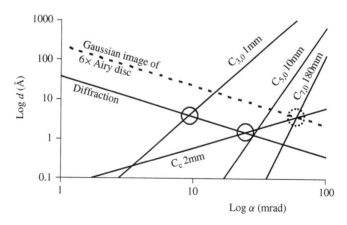

Figure 6.1   Contributions to the probe size on the geometric model. (© Alan Craven.)

expressions for the contributions to the probe diameter due to the source size, diffraction (equation 2.2) and aberrations:

$$d_s^2 = \frac{4I_{\text{probe}}}{\pi^2 B \alpha^2}; \quad d_d^2 = \frac{(1.2\lambda)^2}{\alpha^2}; \quad d_g^2 = A_n^2 C_{n,0}^2 \alpha^{2n} \qquad (6.1)$$

6.2.2 The Minimum Probe Size, the Optimum Angle and the Probe Current

Minimising the sum of three contributions with respect to α gives the optimum probe angle, $\alpha_{\text{opt}}(I_{\text{probe}})$, and the minimum probe size, $d_{\min}(I_{\text{probe}})$, for a given current, I_{probe}, with d_s being set to the required value by choosing the appropriate demagnification of the source. Here $\alpha_{\text{opt}}(I_{\text{probe}})$ is the angle that gives the smallest probe in the presence of a particular set of geometric contributions. The minimisation procedure gives:

$$\alpha_{\text{opt}}^2(I_{\text{probe}}) = \left[\frac{1}{n A_n^2 C_{n,0}^2} \left((1.2\lambda)^2 + \frac{4I}{\pi^2 B^2} \right) \right]^{\frac{1}{n+1}} ;$$

$$d_{\min}^2(I_{\text{probe}}) = \left(\frac{n+1}{n} \right) (n A_n^2 C_{n,0}^2)^{\frac{1}{n+1}} \left((1.2\lambda)^2 + \frac{4I}{\pi^2 B} \right)^{\frac{n}{n+1}} \qquad (6.2)$$

These results show that the probe size depends on the size of the aberration coefficient to the power $1/(n+1)$ but on the wavelength to the power $n/(n+1)$. Thus a much bigger drop in the aberration coefficient is required to reduce the probe size by a given factor than is required in λ. Prior to aberration correction, improved resolution normally relied on going to higher beam energy E_0 in order to reduce λ because only a small reduction in $C_{3,0}$ was possible. However, with aberration correction it is possible to reduce $C_{3,0}$ by a large factor and so improved resolution can be obtained at the same value of λ. This has led to much improved performance at lower accelerating voltages, where knock-on damage is reduced or eliminated (see section 1.5).

The other key conclusion from this analysis is the fact that d_s and d_d depend in the same way on α and enter the results in the same way. Thus the relative size of these two terms determines the current in the probe.

6.2.3 The Probe Current

In the limit of I_{probe} tending to zero, the probe size goes to its absolute minimum size because there is no contribution from the Gaussian image. In practice, there must be some contribution from the Gaussian image in order to have the required current in the probe. As the current is increased by reducing the demagnification, $4I_{probe}/\pi^2 B$ increases relative to $(1.2\lambda)^2$. The probe size increase is small provided $4I_{probe}/\pi^2 B < (1.2\lambda)^2$, e.g. in the geometric approximation, the probe size is increased by 41% when $I_{probe} = \pi^2 B(1.2\lambda)^2/4$. In the non-relativistic regime, B is proportional to the incident beam energy, E_o, while λ^2 is inversely proportional to E_o. Thus the current in the probe is independent of E_o when the Gaussian image is a fixed fraction of the Airy disc size. This conclusion carries over into the relativistic regime.

If $4I_{probe}/\pi^2 B$ dominates $(1.2\lambda)^2$, the balance is not between the aberration and diffraction terms but between the aberration and the Gaussian image terms. If the size of the Gaussian image is a large fraction of the Airy disc, the probe will be incoherent whereas, if it is a small fraction, the probe will be coherent. This has implications for the detection of phase which is discussed in section 6.9.2. Whatever fraction of the Airy disc is chosen for the Gaussian image size, the current in the probe is directly proportional to B so that a high brightness source is essential.

It is clear from the discussion in Chapter 2, section 2.5 that the required brightness is only available from either Schottky emission or cold field emission. While the total current from a Schottky emitter is larger than from a cold field emitter because of the larger source size, the intrinsic brightness of the cold field emitter is higher and its natural energy spread is lower than those of a Schottky emitter. Thus the cold field emitter should perform better at the smallest probe sizes. Having said that, its very small source intrinsic source size means that even more stringent steps are required to prevent interference from increasing the effective source size and hence reducing the effective brightness. Sources of interference include acoustic noise, ac magnetic fields, earth loops and noise on scan and alignment coils. Meticulous attention to detail is required to reduce their effects to a negligible level (e.g. Muller et al., 2006).

A clear advantage of aberration correction is that a very large probe current can be obtained for a modest increase in probe size. The dashed line on Figure 6.1 is where the Gaussian image contribution is 6 times that of the Airy disc. If the third and fifth order contributions are removed by aberration correction, the minimum probe size is obtained at the point where the lines for the Gaussian image, C_C and C_7 meet and the probe size is then similar to the minimum size available in the uncorrected instrument. Compared to the uncorrected case, where the Gaussian image has been set to the same size as the Airy disc, the current has increased by a factor ~ 1000 times since both the Gaussian image size and the aperture size have increased by ~6 times. In practice, aberrations in the gun are likely to come into play reducing the increase in current that is possible and this is a regime where a Schottky emitter may offer better performance because of its larger source size.

6.2.4 A Simple Approximation to Wave Optical Probe Size

As noted above, a wave optical treatment is strictly required to understand the relationships between the various parameters in detail. A simple approach to the estimation of the optimum wave optical performance was given by Conrady (1919)[2], who noted that the rapid increase in the wave aberration with angle in the presence of spherical aberration means that a wave aberration of λ does not alter the diameter of the first zero of the diffraction disc but merely transfers intensity from the central disc to the outer rings. Thus setting $\frac{1}{4}C_{3,0}\alpha_{opt}^{4} = \lambda$ gives $d_{min} = 1.2\lambda/\alpha_{opt}$. This leads to a larger value of α_{opt} and a smaller value of d_{min} than predicted by the geometric model. Assuming this criterion can be applied to higher order aberrations gives a generalised condition for α_{opt}, i.e. $C_{n,0}\alpha_{opt}^{n+1}/(n+1) = \lambda$. In general, the predicted probe angles are always bigger and the probe sizes smaller in this wave optical approximation than in the geometric approximation. Thus the latter must only be taken as a guide on how to optimise performance for a given experiment.

Neither the contribution from the Gaussian image nor that from chromatic aberration are included in the discussion in this paragraph. Wave optical calculations, either full or using the Conrady approximation, deal only with coherent interference effects. The missing contributions involve incoherent effects. Thus the intensity distribution in the Gaussian image has to be convoluted with that from the wave optical calculation while a weighted sum of the intensities from wave optical calculations at different values of defocus is required to deal with chromatic aberration.

6.2.5 The Effect of Chromatic Aberration

When chromatic aberration (equation 2.7) is the limiting aberration, d_{min} is proportional $\sqrt{(C_C \Delta E)}$ so that d_{min} is quite sensitive to both C_C and ΔE, the energy spread of the source. This differs from the

---

[2] Conrady's work on the effect of aberrations on optical telescopes was made use of by M. E. Haine and V. E. Cosslett in their 1961 book 'The Electron Microscope: The Present State of the Art' where limits to resolution and the possibilities for aberration correction are discussed in some detail.

case where a higher order geometric aberration is limiting. Here d_{min} is proportional to the $(n+1)^{th}$ root of the aberration coefficient, $C_{n,0}$, making d_{min} much less sensitive to the actual value of the coefficient. Thus, if chromatic aberration is dominant, it is essential that both C_C and ΔE are minimised. C_C can be minimised by the design of the pole-piece in the objective lens (Krivanek *et al.*, 2003). Prior to the availability of aberration correction, this was not done because it increases $C_{3,0}$. However, this is no longer a concern as $C_{3,0}$ can now be corrected and so the lens can be designed for minimum C_C. ΔE can be reduced by using a monochromator. However, this has the disadvantage of reducing B and hence the current available. Thus a source with the highest value of B and the lowest intrinsic ΔE is required to give the smallest d_{min} with the largest I. Given the recent development of chromatic aberration correction, this situation is likely to change in the not too distant future

6.2.6 Choosing α_{opt} in Practice

As mentioned in Chapter 4 section 4.5.2, the optimum aperture is actually chosen by matching it to the diameter of the central patch of low contrast present in an 'in-focus' Ronchigram from a thin amorphous sample (see Figure 4.12). This patch is often known as the 'sweet spot'. For angles within this sweet spot, the effect of the aberrations on the phase of the ideal wavefront is small and all the rays pass through the same point on the amorphous film giving no contrast in the in-focus Ronchigram. Beyond this angle, the phase shift from geometric aberrations increases rapidly (as α^{n+1} for an nth order aberration). As noted in section 6.2.1, such phase shifts cause deviations of the rays from their ideal trajectories so that they pass through different points on the amorphous film and give rise to contrast. These rays must be excluded by the objective aperture if the probe size is to be as small as possible. Because the phase starts to change rapidly at the edge of the sweet spot, there is a relatively sharp onset of the contrast around the 'sweet spot' and the value of α_{opt} is clear. However, fine control of the aperture size is required in order to obtain the optimum value.

The situation with respect to chromatic aberration is less clear for two reasons. Firstly, the phase change from defocus varies much more slowly over the aperture since it depends only on α^2. Thus the increase in contrast in the Ronchigram as α approaches α_{opt} will be much more gradual. Secondly, a continuous range defocus is present as a result of the energy spread. This will further reduce the contrast in the Ronchigram because it will be averaged over a range of defocus values. Both effects will make it harder to judge α_{opt}.

6.2.7 The Effect of Making a Small Error in the Choice of α_{opt}

At this point, it is worth considering the effect of a small change in α_{opt} on the intensity distribution in the probe. Figure 6.2 is a tableau showing probes formed using two objectives lenses, one uncorrected and one with $C_{3,0}$ set to zero. The calculations use the full wave optical

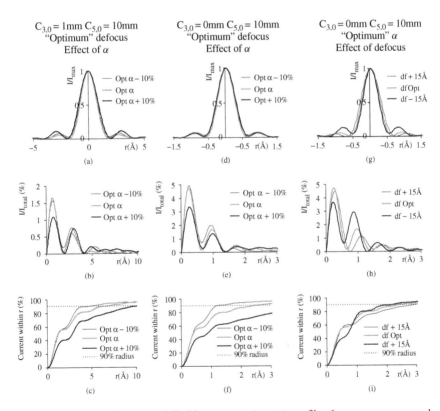

Figure 6.2 Probe shapes modelled by wave optics: (a) profiles from an uncorrected lens showing the effect of probe angle near minimum full width half maximum (FWHM) of central disc; (b) percentage of the current between r and $(r+dr)$; (c) percentage of the current contained within a radius r; (d) – (f) corresponding graphs for a lens with $C_{3,0} = 0$; (g)–(i) the effect of the defocus resulting from chromatic aberration for energy changes typical of cold field emission. (Courtesy of M. Finnie unpublished.)

procedure. The values of α_{opt} were chosen to give the minimum full width half maximum (FWHM) of the central disc, i.e. the best potential image resolution. The top line gives the radial intensity profile of the probe, the second line gives the percentage of the current between r and $(r+dr)$ in the probe and the third line gives the percentage of the current within a radius, r.

Figures 6.2a–c show the effect of a 10% change of α_{opt} on the uncorrected probe. The change in the FWHM of the central maximum is small. Figure 6.2b shows transfer of intensity from the centre to the outer rings and this is most significant as the probe angle is increased. Thus the image resolution is little affected by a 10% change in α but the contrast drops. On the other hand, Figure 6.2c shows that the radius containing 90% of the current drops markedly if the aperture is reduced by 10% and increases even more if it is enlarged by 10%. Figure 6.2c is more relevant when performing analysis (Chapter 7, section 7.8) as the percentage of the current within r controls how far from the centre of the probe a particular signal is being generated, at least in a thin specimen. This highlights the fact that there is no single value of α_{opt} applicable to all situations. What is required is optimisation of the probe for a particular application.

Figures 6.2d–f show the equivalent situations for the corrected lens. Here, the effect of the 10% change in aperture size on the 90% radius in Figure 6.2f is even more marked because of the more rapid variation with angle of the remaining 5th order aberration.

Figures 6.2g–i show the effect of defocus on the probe. The change in focus is chosen to be appropriate for cold field emission with a FWHM of 0.3 eV with an accelerating voltage of 100 keV and a C_C of 1 mm. Again the biggest effect is the transfer of intensity from the centre disc to the outer rings with overfocus having the bigger effect. Because there is a distribution of electron energies present in the beam, the final probe intensity will be a weighted sum of the probe intensities at different defocus values. The phase relationships will be maintained close to the axis but increasingly lost as r increases giving rise to incoherent behaviour. Thus, where C_C is limiting, there is little point in attempting to increase the probe coherence by making the Gaussian image too small since this will result in loss of probe current without an increase in coherence.

6.2.8 The Effect of α On the Diffraction Pattern

Another area of interest is when α is reduced below the value of α_{opt} discussed above. Here the contributions from the aberrations become negligible and the size of the Airy disc increases. By keeping the Gaussian image a fixed proportion of the Airy disc (by reducing the demagnification of the source), I_{probe} can be maintained constant. Thus the angular diameter of the probe can be changed while keeping the area illuminated close to the size of the corresponding Airy disc, giving diffraction information from the smallest possible area. Figure 6.3 shows the effect of the probe angle on the diffraction patterns obtained from [110] zone axis of single crystal Si and from random grains in textured polycrystalline TiN. At a probe half angle of 37mrad, corresponding to an Airy disc

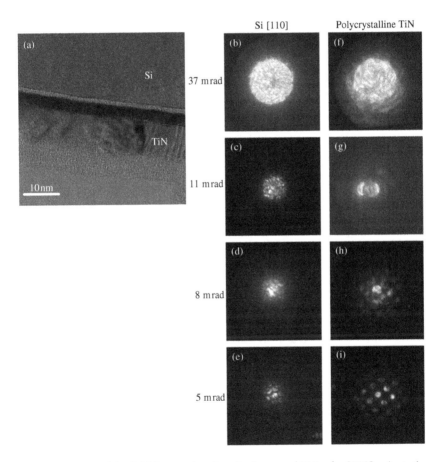

Figure 6.3 (a) Bright field image showing single crystal Si in the [110] orientation and textured polycrystalline TiN with random azimuthal orientations; (b) – (e) diffraction patterns from [110] Si with probe half angles of 37, 11, 8 and 5 mrads respectively; (f)-(i) diffraction patterns from random TiN grains with probe half angles of 37, 11, 8 and 5 mrads respectively. (Courtesy of B. Schaffer unpublished.)

diameter of 0.12 nm at 100 keV, all the discs overlap and there is a mass of detail due to both dynamical diffraction effects in individual discs and interference between overlapping discs. At 11 mrad half angle (Airy disc diameter 0.40 nm), the discs still overlap and it is difficult to tell if the probe is passing through a single grain of TiN. With an 8 mrad half angle (Airy disc diameter 0.55 nm), the diffraction discs from the TiN grain do not overlap and the pattern is essentially from a single grain close to a [100] zone axis. At 5 mrad (Airy disc diameter 0.89 nm), the discs in neither pattern overlap. Thus there are clearly benefits from being able to reduce α when there is a complex microstructure in which more than one crystal can be present in the irradiated volume. Since the probe size here is not limited by the aberrations, this capability is available on any TEM with a bright source and a sufficiently flexible condenser system.

6.2.9 Probe Spreading and Depth of Field

So far, we have considered the size of the incident probe at focus. In vacuum, as the probe propagates away from focus, it spreads. In the geometric limit, it forms a uniformly illuminated disc with diameter $2\alpha z$ where z is the axial distance from the focus to the plane of interest. The smaller the incident probe, the larger α and so the more rapidly the beam spreads. Scattering by the atoms in a specimen will tend to increase the rate at which the probe spreads. Thus as the probe propagates through a specimen of finite thickness, information will come from points in the specimen which are outside the focused probe diameter. However, as discussed in Chapter 5, section 5.4, channelling along atom columns can 'capture' electrons and maintain atomic resolution until the channelled electrons are scattered away. In this situation, atomic resolution information comes only from a layer at the entrance surface and this atomic resolution information sits on a background of lower resolution information generated by scattering from atoms irradiated by the spreading probe.

When the probe angle is large, the probe defocuses rapidly as it propagates. Advantage can be taken of this to give depth sectioning (e.g. van Benthem *et al.*, 2006). Figure 6.4 is an example of a through focal series of a Si nanowire containing single Au atoms showing them becoming sharp when the probe is focused at their depth. There is also an increasing interest taking advantage of this depth sensitivity by developing confocal methods for the extraction of three dimensional images (e.g. d'Alfonso *et al.*, 2008).

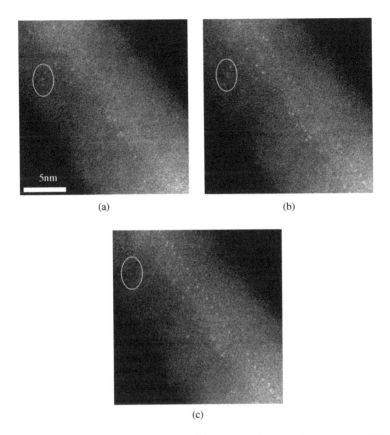

(a)

(b)

(c)

Figure 6.4 High angle annular dark field images of single Au atoms at different depths in a Si nanowire. The focus of the probe is changed by 6 nm along the optic axis from (a) to (b) and again from (b) to (c). The group of atoms in the ellipse is in focus in (a), out of focus in (b) and almost invisible in (c). There is a set of atoms down the centre of the wire that are close to focus in (b) and a further set to the right which are close to focus in (c). (© Alan Craven.)

When using an analytical technique such as EELS or EDX (Chapter 7), the signal comes from all the atoms that are irradiated by the primary beam. For example, if the probe is focused on the entrance surface of a specimen of thickness, t, as might happen when using a HAADF imaging technique, geometrically the probe will have a diameter $2\alpha t$ at the exit surface. If the problem being investigated is diffusion at an interface parallel to the incident direction, then this beam spreading will limit how well the extent of the diffusion can be determined. A first guess at the best set-up would that would minimise the beam spreading would be to set the probe size at the entrance surface equal to the geometric spread at the exit surface, i.e. make $1.2\lambda/\alpha = 2\alpha t$ so that $\alpha^2 = 0.6\lambda/t$. This gives a central disc with a diameter of $\sqrt{(2.4t\lambda)}$ which should vary little as it passes through the specimen. Focusing the probe at the centre of the specimen would reduce it further While the optimum conditions remain to be determined exactly, the underlying arguments are clear. It is also clear that the best spatial resolution requires very thin specimens but they must also be free of artifacts, something requiring significant effort.

6.3 THE CONDENSER SYSTEM

In a microscope column incorporating an aberration corrector, such as described in Chapter 4, the role of the condenser system is to take the cross-over provided by the gun and

- form a Gaussian image of it of the size required at the specimen;
- create the desired probe half angle, α;
- maintain an on-axis intermediate image of the gun cross-over at a fixed plane in front of the corrector.

The need for the first two features is clear from the discussion in sections 6.2.3 and 6.2.6, where it was noted that very fine control of α is required to optimise the probe. Such fine control is not possible by simply using a set of interchangeable physical apertures with different but fixed sizes. Hence the condenser optics must be used to provide the necessary finesse in the control of α. In this way, a single physical aperture in the column can act as a virtual aperture giving a wide range of α at the specimen. It is also possible to use alignment coils to position the aperture rather than mechanical motion.

The third feature above is required so that the corrector does not need major retuning when d_s and/or α are changed. If the source is always on-axis at a fixed point, the first order trajectories (see section 4.2) through the corrector remain the same and the only re-tuning required is similar to that which is needed periodically during the working day.

To allow these three features to be set independently requires three degrees of freedom and so the condenser system must contain a minimum of three lenses whose strengths can be varied independently (e.g. Krivanek *et al.*, 2008b). A system which demonstrates how the required feature might be obtained is shown schematically in Figure 6.5. Since the aim is to keep first order optical properties of the corrector and the objective lens fixed, the intermediate cross-over will subsequently be

Source C_1 Ap C_2 C_3 Image for
from gun   corrector

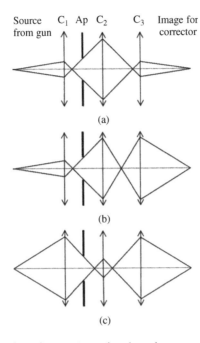

(a)

(b)

(c)

Figure 6.5 Some modes of operation of a three lens condenser system (C1, C2, C3) with fixed planes for crossovers from the gun and for the corrector and a single physical aperture (Ap). (a) and (b) have the same angle of acceptance from the gun and hence the same I but different α and hence different d_s; (b) and (c) have the same α but the angular magnification from source to image differs and so they have different d_s and I; (a) and (c) have different α but the same angular magnification and hence the same d_s but different I. (© Alan Craven.)

demagnified by a fixed amount whatever the settings of the condensers lenses. Thus the properties of the probe will directly reflect the properties of this intermediate cross-over.

Figures 6.5a and 6.5b show a situation that will result in a probe with the same current, I, because the same solid angle is accepted from the gun. However, there are different values of the convergence semi-angle, α. Since the angular magnification is greater in Figure 6.5b, the Gaussian image, d_s, is smaller. This follows since the linear demagnification is proportional to the angular magnification. Figures 6.5b and 6.5c will give the same value of α but different values of d_s because the angular magnification differs. Furthermore, the set-up in Figure 6.5b gives a lower current than that in Figure 6.5c because it accepts a smaller solid angle from the gun. Figures 6.5a and 6.5c have the same angular magnification and hence will give the same d_s but different values of α and hence I_{probe}.

In simplistic terms, the C_1 lens controls the current in the probe and a combination of C_2 and C_3 define the probe angle and the spot size.

6.4 THE SCANNING SYSTEM

6.4.1 Principles of the Scanning System

The reciprocity theorem was introduced in Chapter 3, section 3.3. Figure 6.6a shows the ray diagram of conventional CTEM imaging two points on a specimen which is illuminated from a point source. The CTEM has an aperture in the back focal plane of the lens. To get the equivalent performance from the lens in STEM mode, the probe must be placed successively at each conventional image point so that the rays from it follow the rays in the conventional CTEM image but in the opposite direction, as shown in Figure 6.6b. Thus the central ray of the probe must pass through the focus of the lens at each probe position, as shown in Figure 6.6b. In this way, the central ray of the probe is parallel to the axis when it arrives at the specimen. This motion of the probe can be achieved from a single source of electrons by using a set of double deflection coils before the specimen as shown in Figure 6.6c. By choosing the ratio of the deflections in the upper and lower coils, the axial ray is made to re-cross the axis at a chosen point. This point is often called the rocking point. Here the rocking point should be at the focal point of the lens. This is equivalent to setting up pure shift of the illumination in a CTEM. While Figures 6.6b and 6.6c show the limiting aperture next to the objective lens, in practice, the real aperture is in the condenser system before the corrector as discussed in section 6.3.

The probe in STEM generates a diffraction pattern on the plane where the detector is placed. Correct setting of the rocking points makes the diffraction pattern stationary on the detector as the probe is scanned. While the diffraction pattern may be stationary, Figure 6.6b shows that a ray arriving at a point on the detector changes direction as the probe scans. Thus to ensure that the central ray travels down the axis of subsequent optical components, the probe should be 'descanned' using double deflection coils after the specimen. Lack of descanning can cause a systematic shift of the energy loss spectrum when the probe is scanned along the direction that maps onto the dispersion direction of the spectrometer (Chapter 7, section 7.2.2). It can also cause a loss of intensity in the spectrum when scanned in the perpendicular direction.

While having a rocking point at the focus works well for high magnification, there will be a maximum current that can be passed

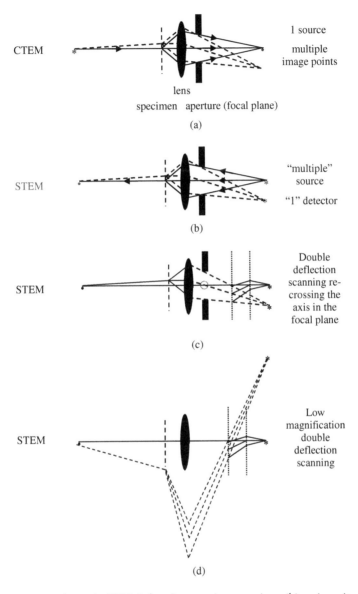

CTEM

1 source

multiple
image points

lens
specimen   aperture (focal plane)

(a)

STEM

"multiple"
source

"1" detector

(b)

STEM

Double
deflection
scanning re-
crossing the
axis in the
focal plane

(c)

STEM

Low
magnification
double
deflection
scanning

(d)

Figure 6.6 (a) schematic CTEM showing two image points; (b) reciprocity equivalent STEM showing the source positions required to image the same two points;
(c) the double deflection system which allows the use of a single source to achieve
the result in (b) with the rocking point marked with the dotted circle; (d) the
double deflection system set up to give much larger probe motion to provide low
magnification. (© Alan Craven.)

through the scan coils. This limits the probe deflection and hence sets a limit to the lowest magnification that can be achieved, which is typically in the range 1000–10 000 times. One way to achieve a lower magnification is to make both deflection coils work in the same sense. In this way, the probe moves across a large region of the specimen, giving low magnifications as shown in Figure 6.6d. The limit is normally set by the bore of the objective lens or the specimen holder. However, in this mode the diffraction pattern scans over a large distance in the detector plane unless there is effective descanning. This can cause changes of contrast and may even convert bright field contrast to dark field contrast if the cone of illumination from the probe moves off the detector. Even without descanning, this low magnification setup is a very useful 'search' mode for finding features. Having said that, low magnification information is also available in the Ronchigram at large probe defocus (see the central region of the Ronchigram in Figure 4.12).

6.4.2 Implementation of the Scanning System

Magnetic deflection of the beam is almost invariably used for the scan. The traditional arrangement is a raster scan composed of a fast 'sawtooth' ramp for the line scan and a slow 'sawtooth' scan in the perpendicular direction for the frame. All modern scan generators produce the sawtooth ramp signals for the raster digitally and these signals pass through digital to analogue converters (DACs) to produce the analogue currents that pass through the scan coils. The electronics allows rotation of the raster pattern and a shift of the raster pattern. The latter provides electrical shift of the image and needs to be independent of the size of the raster so that the image does not displace sideways as the magnification is changed. All scan generators have a range of features including: a reduced raster (to allow focusing/stigmation), a line scan (for line traces) and a point mode (for point analysis). All allow the dwell time per pixel to be varied over a wide range.

6.4.3 Deviations of the Scanning System From Ideality

Modern electronics is so fast that the limit for the minimum practical dwell time per pixel is typically limited by two other effects. One is the phase shift caused by the eddy currents induced in the conducting vacuum wall of the flight tube that passes through the scan coils. For dwell times below a critical value, this phase shift is sufficient to displace the image in the line direction by an amount that increases rapidly with decreasing dwell time. Unless the dwell time is longer than this critical dwell time, there can be positioning errors when using the image to position line scans or stationary probes. The second limiting factor is the flyback time of the line scan. When the scan reaches its maximum value, it needs to return to its minimum value as quickly as possible. However, the inductance of the scan coils combined with the limited supply voltage in the electronics sets a limit to how fast this can occur and the trajectory of the beam tends to be uncontrolled during such flyback. Thus a suitable delay period must be allowed for the beam to settle before the scan starts again. If this delay is too short, one side of the image becomes distorted. However, if it is too long, additional damage may occur during this settling time. These effects are often 'hidden' from the user by including appropriate delays in the control software.

In addition to the intentional deflection of the probe by the scan generator, there can be a deflection caused by magnetic fields from other sources, so-called stray fields. These are often related to the electrical supply and appear at the fundamental frequency of the supply and its harmonics. There is a wide range of ways in which the electrical supply can introduce such stray fields (Muller et al., 2006) and it is essential that they are tracked down and eliminated if the best performance is to be achieved. However, their effects can be 'hidden' if the start of a scan is synchronised with a particular point in the supply waveform. In this way, the effect of the interference is locked to the raster so that the probe follows a well defined but distorted path in each line. This distorts the image but maintains the resolution as shown in Figure 6.7a where the line scan starts a fixed point in the supply voltage waveform. Figure 6.7c shows that sometimes the deflection from the stray source adds to that from the line scan giving a bigger rate of change of position and hence lower local magnification and sometimes it subtracts from it giving locally higher magnification. However, the resolution is maintained locally. Of course, the deflection from the stray source can have a complex wave form and be both along and normal to the line scan direction. Figure 6.7b show the result when the stray deflection is not synchronised to the supply frequency. Here, each successive line is displaced 'randomly' in position and so there is no average distortion but there is a loss of resolution. This behaviour will occur for any stray deflection not synchronised to the line scan. Such deflections need not be magnetic in origin but can come from a range of effects, e.g. vibration.

If imaging is the sole aim of the work, synchronisation successfully maintains resolution at the expense of distortion. However, if the probe

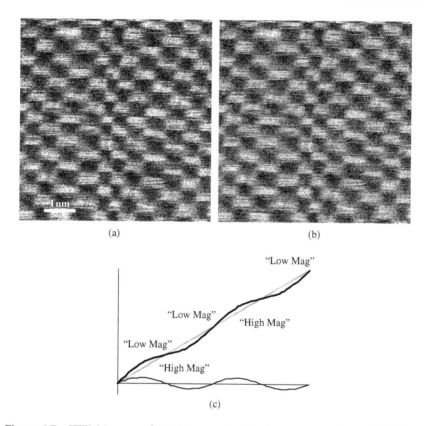

(a)   (b)

(c)

Figure 6.7 STEM images of [110] Si recorded in the presence of stray field from the electrical supply. (a) the start of the line scan is synchronised to the supply frequency showing regions of high and low effective magnification; (b) a similar image without synchronisation; (c) a schematic diagram of the ideal scan deflection (grey), the deflection by the stray field (thin black) and the overall deflection (thick black). Where the slope is low the magnification is high and vice versa. (© Alan Craven.)

is stopped for point analysis, the source of stray deflection will continue to move the probe on the specimen and so increase the area irradiated. Thus for high resolution analysis requiring integration over more than the period of the interference, it is essential to eliminate the source.

6.4.4 The Relationship Between Pixel Size and Probe Size

With a flexible, low-noise scanning system in place, there are three length scales to consider. The size of the probe itself, the pixel size and the feature size in the specimen, e.g. the separation of the atomic columns.

With a digital scan generator, the pixel size is just the step size in the raster. If this is smaller than the probe size, the image is over-sampled and information from a given point on the specimen appears in more than one pixel. If the pixel size is bigger than the probe size, the image is under-sampled and some regions of the specimen do not contribute to the image. One way to overcome this is to defocus the probe to match the pixel size. This has the advantage of reducing the instantaneous current density but is difficult to control accurately as the distribution of the current in the probe changes significantly with defocus, as can be seen in Figure 6.2 g to i. Alternatively, the scan system itself can scan the probe within the pixel in a process known as sub-pixel scanning. This process is very controllable and reduces the average if not the instantaneous current density.

In a periodic system, if the pixel size is not an integer fraction of the periodicity of the specimen, there will be an aliasing effect whereby, when the probe falls on the atom centre in one pixel, it will not fall on the centre of the neighbouring atom. Thus the image will not truly represent the sample. The situation is clearly more complex in aperiodic systems and the full analysis involves the Nyquist sampling theorem. In essence, a system must be sampled with a period half that of the maximum frequency of interest, i.e. at least the maximum and minimum (antinodes) of the waveform need to be sampled to see that there is a period. In reality, things are a little more complicated, e.g. what happens if you sample nodes of a periodic system? A useful introductory discussion can be found in the book by Press et al. (1992).

6.4.5 Drift, Drift Correction and Smart Acquisition

Since STEM records the pixels sequentially, drift causes distortion of the image but typically does not cause significant loss of spatial resolution for a given pixel. For long exposures, such distortion can be minimised by using drift correction. A drift correction area with appropriate contrast and detail close to that under study is selected and a short exposure survey image including it is recorded. Figure 6.8a shows an HAADF survey image identifying the regions used for a spectrum image and for drift correction. The short time taken to record this means that drift has a negligible effect. However, in the much longer time required for the spectrum image, the drift is significant. In this case, the spectrum image acquisition is interrupted after every tenth line and the image from the drift correction region re-acquired. Comparing this to the original allows the magnitude and direction of the drift to be determined and

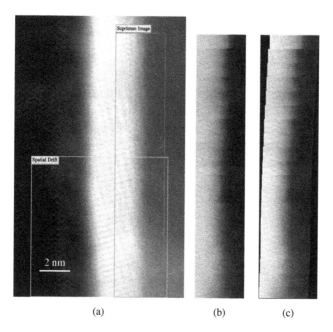

Spectrum Image

Spatial Drift

2 nm

(a)   (b)   (c)

Figure 6.8 (a) An HAADF survey image showing the regions selected for a spectrum image and for drift correction; (b) The HAADF signal recorded as part of the spectrum image with drift correction; (c) The image in (b) with the drift correction offsets removed (Courtesy of B. Schaffer unpublished.)

corrected. Figure 6.8b shows the image with the drift correction applied and there are clear jumps because the drift continues steadily between the corrections. With the contrast available in this image, further correction by 'eye' is possible but in other cases it might not be. Figure 6.8c has the drift correction 'removed' by displacing the blocks of ten lines in the opposite sense to the correction applied to them, giving an indication of the level of the drift occurring.

The flexibility of the scan system also allows the possibility of so called *smart acquisition* (e.g. Sader *et al.*, 2010). In principle, from a rapidly acquired image of an area, the positions and shapes of features of interest can be identified using image analysis techniques. These can be used to construct image masks. Data can then be collected only from those regions of interest. Drift correction makes sure the features are always correctly located. This is useful for gathering data from boundaries, edges of particles or a set of very small randomly distributed particles. It will be especially useful where there is a set of features which are individually too radiation sensitive to give the required spectral data. Using smart acquisition, a spectrum can be obtained by integrating the signal from a large number of the features, each of which receives only a low electron fluence (see also Chapters 7 and 8). With control of other features of the microscope, such as the stage, automated data acquisition over long periods of time can potentially be implemented allowing unattended overnight acquisition of many datasets.

6.5 THE SPECIMEN STAGE

The specimen stage is required to move the region of interest in the specimen to the right point on the axis at the required orientation to the beam without introducing vibration or drift. There is often interest in heating or cooling the specimen as well as a range of more specialised requirements, e.g. applying a controlled atmosphere or applying electric or magnetic fields. For an aberration corrected system working at better than atomic resolution, very stringent requirements on noise, drift and positioning are required. Not only accurate transverse positioning is required but also accurate height positioning. If the height change exceeds a critical value, the aberration corrector for the objective lens needs major retuning as the first order trajectories change. Keeping the specimen at constant height also keeps the objective lens excitation constant and hence the post-specimen optics unchanged. Thus, as the transverse position is changed and as the tilt is changed, the height of the specimen must be kept constant.

The eucentric 'side-entry' rod has advantages for maintaining height on the two transverse axes and one tilt axis but needs correction for the second tilt axis. Thus it has many practical advantages. However, the asymmetry of the construction can render it more susceptible to drift and vibration. The 'top-entry' entry system can be much less sensitive to drift and vibration since a cartridge is placed in a stage sliding on the objective lens pole piece in an arrangement with something closer to cylindrical symmetry. However, now all transverse motion and tilt requires correction for the change in specimen height they cause. With sufficiently accurate manufacture, both the side-entry stage and the top-entry stage can use computer control to achieve a reasonably constant height as position and tilt are changed.

Whichever system is used, the environment and the services are critical to achieving the required performance (Muller et al., 2006). The air and water temperatures must have very slow rates of change (ideally <0.1°C per hour). It is crucial to realise that this is a specification on the rate of change of temperature not the maximum temperature change, i.e.

it is possible that the temperature always lies in the range 20°C to 20.1°C but that it swings relatively rapidly between the extremes giving an unacceptable rate of change of temperature. An example is shown in Figure 6.9. The turbulence and pump induced pressure changes in the cooling water should also be sufficiently low that they cause no vibration. The acoustic noise levels in the room should be sufficiently low that they have no effect. Here suitable acoustic damping is valuable. The mechanical couplings into the column should also be designed so they are not sensitive to the room pressure. The microscope itself should be mounted on a floor that has low vibration and have a suitable vibration isolation system. To help to maintain stability at the highest possible level, constant power consumption should be maintained in the column. Modern instruments can have multiple coils in the lenses so that the lens excitation can be varied while keeping the power dissipation constant. There is some advantage in having remote operation so that the operators themselves have no perturbing effect on the column. Alternatively, the column itself can be isolated within a suitable box which provides mechanical, thermal and magnetic shielding.

Traditionally, high resolution has only been obtained by keeping the sizes of the bore and gap of the objective lens small so that the spherical aberration is kept small. With the advent of aberration correction, this

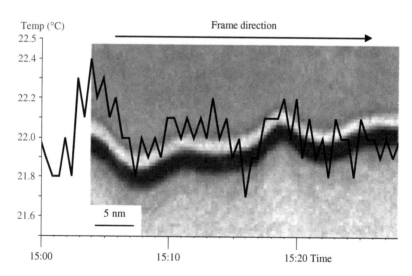

Figure 6.9 Drift due to slow variation in the water temperature from a chiller. The temperature of the chiller output is superimposed on the image so that the time scales match. On a fast scan the interfaces are straight. (Courtesy of S. McFadzean unpublished.)

constraint is removed. There is the opportunity to increase the space in the objective region to allow more room for specimen manipulation and a wider range of ancillary techniques. There is also the possibility of providing controlled atmospheres for the specimen (e.g. Gai and Boyes, 2009).

6.6 POST-SPECIMEN OPTICS

So far in this chapter, we have considered the formation and scanning of the probe on the specimen in a way consistent with sub-atomic resolution. Equally important is the efficient detection and recording of the available signals. The detectors we use for high energy electrons have fixed physical size and we need to match the size of the electron distribution emerging from the specimen to the detector size to be used. As introduced in Chapters 2 and 3, the easiest way to do this is to use post-specimen optics. In STEM, the spatial resolution depends on the probe size and normally a diffraction pattern is present on the detectors. To make best use of the diffraction pattern, the ability to change the camera length is important (Chapter 7, section 7.2.2). Control over the camera length also plays a key role in electron spectrometry where the angular acceptance of the spectrometer is often much smaller than the angular range which needs to be collected. Thus a short camera length is required compress the angular distribution. To do this successfully, it is important to compress the scattered electron distribution before it leaves the objective lens so that the distribution enters the next lens sufficiently close to the axis. If this is not achieved, the aberrations of this subsequent lens will limit performance. With the 'condenser-objective' type lens, where the focusing action after the specimen is similar to that before the specimen, compression within the objective lens is easily achieved. The overall increase in column length introduced by the post-specimen optics can be minimised if non-round optical elements are used, e.g. as in the quadrupole coupling module in the SuperSTEM instruments (Krivanek *et al.*, 2008b). These have an added advantage in that they can also be used to correct some of the higher order aberrations in the EELS spectrometer.

As in a CTEM column, post-specimen lenses introduce distortions and chromatic effects. For a cylindrically symmetric detector like an annular dark field detector, it is important that the radial distortion is kept low if there is to be a linear relationship between angle of scattering and position on the detector. This is normally assumed but can be hard to verify in practice. Where non-cylindrically symmetric detectors are involved, spiral distortion must be kept low. An example would be a quadrant detector where a circular detector is split into four independent quadrants allowing the magnitude and direction of the deflection of the scattering distribution to be determined. The optical arrangements for achieving low distortion in post-specimen lenses are well known in the CTEM.

Aberration correction of the probe has increased the challenge of signal collection for EELS because the probe convergence semi-angle is increased very significantly (e.g. ~50 mrad probe half angle for 0.6 Å probe diameter at 100 keV). It is essential to collect over at least this angular range for relatively low energy losses and over an even larger range for higher losses, otherwise a large fraction of the available signal is lost (see section 7.8). At the compressions needed for such collection angles, there can be a considerable variation of the camera length with energy loss and the detailed set-up of the optics is quite important. Figure 6.10a shows that if real images of the probe are formed by the post-specimen optics, the camera length always increases with energy loss so that the collection angle decreases. However, if lenses P_1 and P_2 do not produce real crossovers, the camera length can be made to decrease with energy loss, as shown in Figure 6.10b (Craven and Buggy, 1981). Thus it should be possible to make it independent of energy loss. In addition, the cross-over after P_2, which is the source for the spectrometer, moves towards P_2 with increasing energy loss. This is a smaller effect for the real image case than for the virtual image case. This shift of the source for the spectrometer leads to a defocus of the spectrum with energy loss. In principle, this can be corrected by tilting the dispersion plane of the spectrometer with suitable correcting elements in the spectrometer. However, correction of chromatic aberration in the post-specimen optics will undoubtedly be the best solution.

6.7 BEAM BLANKING

There are two basic uses of beam blanking in the STEM. One is the removal of the probe from the specimen when signals are not being recorded. This limits the fluence given to the specimen and is extremely valuable if the specimen is susceptible to radiation damage. However, noise in such pre-specimen blanking has the potential to introduce unwanted deflection of the probe. In addition, time variation of the local potentials will occur on insulating specimens and can lead to artifacts. The second use is to provide a shutter for detectors, particularly detector arrays such as a CCD camera. In the CCD cameras typically used in

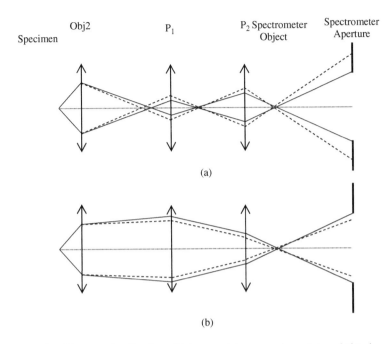

Figure 6.10 Change of collection efficiency with energy loss. (a) and (b) show the same collection half-angle for the elastically scattered electrons (solid lines). The lenses become stronger for those electrons which have lost energy (dashed lines). The configuration in (a) collects a smaller angular range of the loss electrons than that for the elastic electrons whereas that in (b) collects a larger angular range. Note that the cross-over which acts as the spectrometer object moves towards P2 in both configurations as the lenses are assumed to be round lenses. (© Alan Craven.)

TEM, the signal needs to be removed from the detector during read-out to avoid smearing when the charge from a specific pixel is stepped across the array to be read out. Beam blanking may also be used to prevent the saturation of the detector in an electron spectrometer and to improve the throughput of EDX detectors, as discussed in Chapter 7, section 7.2.3.2. Electromagnetic blanking is used for slow shuttering but to achieve the rapid settling times needed for beam blanking and fast shuttering, electrostatic deflection is essential (Craven *et al.*, 2002).

6.8 DETECTORS

6.8.1 Basic Properties of a Detector

At the start of this chapter, it is said that an ideal detector adds no noise to the signal. In reality, all detectors add some noise. The art is to keep the noise added to a very low level. The standard way of describing the noise added by a detector is the detective quantum efficiency (DQE). The DQE is the ratio of the square of the signal to noise ratio (SNR) at the output of the detector to the square of the SNR in the input signal, i.e. $(SNR_o/SNR_i)^2$. Thus an ideal detector has a DQE of 1 while a real detector has a DQE < 1. One way to think about this is that the square of the SNR in a Poisson distribution is just the number of electrons collected. If n electrons arrive and the detector has a DQE of 0.1, the effective number of electrons detected drops to $0.1n$ because the $(SNR_o)^2$ has dropped by a factor 10. This immediately gives a clue to the key property required in a detector with high DQE, i.e. the signal created by an incident electron must remain above the noise level at all points in the system otherwise its arrival will not be noticed at the output and the effective number of electrons recognised will therefore drop. This means that most practical detection systems have in-built gain to ensure a large response to each incident electron. This is perfectly feasible for high energy electrons and high energy photons, possible for secondary electrons, which can be accelerated before the detector, but difficult for optical photons and induced specimen current, unless generated in the depletion region of a pn junction where the electric field separates the electrons and holes before they have chance to recombine.

Most detection systems produce an output in the absence of a real signal. In electronic systems this is known as dark current and corresponds to the fog on a photographic plate. This will have an average component, which can be subtracted off, and a noise component, which cannot, e.g. read-out noise. All detection systems have a maximum output signal known as saturation. Thus a given detector has a dynamic range, often defined as the saturation signal/read-out noise. If the dynamic range of a detector is low, there will be problems recording very high contrast images, diffraction patterns and electron spectra. It must be remembered that, while the dynamic range is a good indicator of performance, the minimum intensity in an image or spectrum must be much higher than the read-out noise level and so the useful intensity range of a detector will always be less than the dynamic range defined above.

Between the dark current and the saturation output, the output signal should be proportional to the input signal, i.e. the detector should be linear in its response. Deviations from linearity often occur as the saturation signal is approached, e.g. in pulse counting systems, pulses start to overlap and so some are missed, an effect known as deadtime. Detectors also have a limited bandwidth, i.e. a maximum frequency response, f_{max}. This will determine the minimum usable dwell time (typically taken as $1/(2f_{max})$) and hence the minimum time to record a frame. Unless this is sufficiently short, operation of the instrument can become tedious. A TV rate frame makes searching and set-up easier even if the SNR is poor but this requires a $\sim$5 MHz bandwidth. Ten frames per second is a reasonably comfortable rate and this frame rate can be obtained for a frame with 256×256 pixels using a 2 μsec dwell time. However, if the frame rate drops below 1 frame per second, operation (as opposed to recording) becomes tedious.

6.8.2 Single and Array Detectors

For scanning systems, there are two basic forms of detector, a single detector where the signal is combined into one output and an array (or configured) detector where electrons arriving in different regions give rise to different signals. The latter group divides into ones where the separate signals are available continuously, e.g. a quadrant detector, and ones where there is an integration period followed by a read-out period, e.g. a CCD camera.

All modern detectors give a digital output. Some are essentially analogue with an analogue to digital converter (ADC) at the output, e.g. a CCD camera, while others detect and count the pulses created by individual electrons arriving at the detector. Some can operate in both modes, e.g. scintillator/photomultiplier systems can operate either in pulse counting mode or analogue mode (or both). A typical EDX detector combines both digital and analogue processing in that each X-ray is detected as a pulse but the charge in the pulse is measured to give the X-ray energy.

6.8.3 Scintillator/Photomultiplier Detector

The scintillator/photomultiplier combination is an excellent single detector giving a high DQE, a high dynamic range and a high bandwidth. A scintillator, e.g. cerium -doped yttrium aluminium garnet (YAG), converts about 1% of the electron energy to visible photons, i.e. a 100 keV incident electron generates $\sim$300 photons. The principal decay constant of YAG is $\sim$100 nanoseconds and so the pulse of photons is produced in a similar time. Note that YAG also has much longer lived metastable states which can store a small amount of energy for hours. Thus there can be a problem with increased dark current if a high intensity has been incident on the detector. This point is revisited in the discussion of CCD cameras in section 6.8.5 ahead.

The photocathode of the photomultiplier has a quantum efficiency of ~20% so that a photon incident on the photocathode has a ~20% chance of ejecting a photoelectron. Thus ~300 photons should give ~60 photoelectrons. These photoelectrons are accelerated to a dynode where each generates secondary electrons which in turn are accelerated into the next dynode. In this way, a gain of 10^7 can be obtained and the gain can be varied simply by changing the voltage across the dynodes. The name of the control which does this is typically 'gain' or 'contrast' depending on the manufacturer. Transit through the photomultiplier is so fast that the width of the output pulse is controlled by the decay time of the YAG. The final electron pulse is collected on the anode which has very low values of both output impedance and capacitance and so is well suited to be the input to further fast amplification.

If all the photons from the scintillator reach the photocathode, the signal from an incident electron is much bigger than a dark current pulse resulting from a single photoelectron emitted from the photocathode by thermionic emission. In pulse counting mode, such small pulses can be discriminated against so that the DQE of the scintillator/photomultiplier is essentially 1. Pulse pile up problems start to occur at count rates of a few MHz so that the maximum input signal in this mode is a few tenths of a picoamp. For currents of this level and higher, the fraction of the output current that is from the dark current is negligible and analogue amplification of the output followed by an ADC also gives a very high DQE.

In reality, it is impossible to get all the photons to the photocathode and surprisingly easy to get very few there. The trick is to get enough to ensure that the signal pulse is sufficiently bigger than the dark current pulse that DQE is not decreased significantly. The issues are that the photomultiplier tube and its housing are bulky and so must be outside the column. At the same time the photons are emitted isotropically in a medium of relatively high refractive index. Coupling via a light pipe transfers a significant fraction of the photons provided optical coupling is used at each interface, something not always desirable from a vacuum point of view. Figure 6.11 shows a schematic diagram of an annular detector using a light pipe. Alternatively, carefully designed mirrors or lenses, while not as effective, can still transfer sufficient photons to make the signal pulses bigger than the dark current pulses.

The signal detected using a scintillator/photomultiplier detector, like that from any other detector, may have a component that changes slowly as the probe is scanned across the sample and another that changes rapidly. There is normally the provision to boost the contrast by subtracting a DC level (using the 'black level' or 'brightness' control)

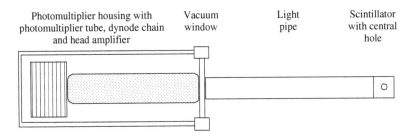

Figure 6.11 Schematic of an annular detector using a light pipe coupled scintillator/photomultiplier combination. (© Alan Craven.)

to null the slowly varying component. Adjusting the gain (contrast) and black level (brightness) controls can increase the contrast of the image on the display but cannot increase the SNR, since both the signal and the noise are amplified. However, when absolute signal measurement is required, care must be taken to adjust the black level (brightness) setting so that zero output signal corresponds to zero input signal.

Overall, the scintillator/photomultiplier combination is an 'ideal' single channel detector. All the image detectors on the SuperSTEMs use this type of detector.

6.8.4 Semiconductor Detectors

Gain is also available in the depletion layer of a semiconductor device. For every ~3 eV of energy deposited in the depletion layer of a Si device, an electron-hole pair is created and one unit of charge can be made to flow through an external circuit. Thus, if a 100 keV electron is stopped in the depletion layer, ~30 000 electrons flow through the external circuit giving a high gain. One problem is that the range of 100 keV electrons in Si is somewhat larger than the width of the depletion layer in an unbiased Si *pn* junction. The range can be reduced by going to a semiconductor of higher atomic number, e.g. GaAs. The depletion layer width can be increased by applying a reverse bias but, in Si at room temperature, there is a large contribution to the noise from the resulting noise in the reverse leakage current. The depletion layer width can also be increased by using lightly doped material. Most semiconductor detectors use the latter strategy and work at room temperature with zero bias since the added complexity of a cooling system to reduce the reverse current often removes the advantages of the simplicity and small size of the detector elements. Thus, while the maximum gain may not be achieved, the actual gain is still large.

Semiconductor detectors are also used for energy dispersive X-ray detectors because an individual x-ray photon produces a large number of electron holes pairs. Provided that all the energy of the photon is absorbed in the depletion layer, the average number of electron hole pairs created is accurately proportional to the photon energy. The charge generated is integrated on a capacitor at the input of a field effect transistor and the resulting change in the voltage at the output is digitised to give a measure of the pulse height and hence the photon energy. Noise in the electronics sets a lower limit to photon energy that can be determined this way. In the traditional Si(Li) detector, a piece of intrinsic silicon, typically 3 mm thick and with an area of tens of square millimetres, has p- and n-junctions formed on opposite faces. By cooling close to liquid nitrogen temperature, the whole volume can be depleted with negligible leakage current, giving high efficiency detection for photons that are fully absorbed within the volume. As the photon energy exceeds $\sim$15 keV, there is an increasing probability of the photon being transmitted through 3 mm of Si without absorption and so the detection efficiency starts to drop. At low energies, absorption of the photons before they enter the depletion layer limits detection efficiency. This is dealt with more fully in Chapter 7, section 7.2.1.

Semiconductor detectors have the great advantage that they are straightforward to make either as Schottky barrier diodes or as surface diffused pn-junctions. Their size and shape can be simply controlled and thus it possible to make configured detectors with a number of elements, e.g. a quadrant detector. Semiconductor detectors are small and easy to mount in the column, needing only electrical feedthroughs. Their disadvantage is that they have a large capacitance and this leads to large gain for high frequency noise in the subsequent amplification. Thus the bandwidth must be limited, typically to within the range 10–100 kHz, giving frame rates of only a few per second. Control of the signal level is provided by changing the gain of an amplifier after the detector and black level (brightness) is introduced by adding an offset voltage at that stage.

6.8.5 CCD Cameras

The single detectors described above are excellent for recording images from specific signals. However, it is also necessary to view and record diffraction patterns, Ronchigrams and even the CTEM images that are available with suitable post-specimen lenses. Having access to a fast read-out of the Ronchigram is essential to tuning the corrector. By far the most widely used type of camera is currently based on the charge coupled device (CCD).

A CCD is a regular array of cells, each of which is a metal oxide semiconductor capacitor. The gate voltage applied to a single capacitor depletes the carriers below it forming a potential well. As electron hole pairs are created, carriers become trapped in the cell and so a signal can be integrated. After a suitable integration time, the input is removed and the stored charges read out. There are various read-out schemes for CCDs but they all use a method in which clock signals change the potentials between adjacent cells in such a way that the stored charge is moved from cell to cell with essentially no loss. In one arrangement, there is an output register at one side. This is just a line of cells matching the last line of cells in the active array. Sometimes the output cells have a larger capacity to allow the contents of several array cells to be added together, a process known as binning. The first step in the read-out is that the charges in the rows of cells parallel to the output register are clocked one step sideways so that the contents of the first row of active cells is moved into the output register. The charges in the cells in the output register are then clocked along the output register to the output amplifier whose output is digitised to give the signal from each cell in turn. When this is complete, the rows of active cells are clocked one more step towards the output register so that the second row is now in the output register. This process continues until the whole array is read out.

The time to read out the array depends on the number of cells in it. Large CCDs can be divided into sections, each with their own output register and output amplifier, allowing these sections to be read out in parallel. The speed also depends on the accuracy with which the signal is to be digitised. 14-bit accuracy takes much longer than 8-bit accuracy. A 1k × 1k array digitised to 14-bits typically takes 500 msec to read out but TV rates are possible at lower accuracy.

It is possible to directly irradiate a CCD with 100 keV electrons but the CMOS circuitry undergoes radiation damage with the result that the device degrades rapidly. The large number of carriers created per incident electron also means a cell is filled by a small number of incident electrons giving a very low dynamic range but an excellent DQE. Almost all CCD cameras used in TEM convert the incident electrons to optical photons in a scintillator (or phosphor) as shown in Figure 6.12. The scintillator is normally coupled to the CCD using one or more fibre optic plates but lens coupling is also used. Typically one fibre optic plate is bonded to the scintillator, whose other face has a thin metal coating to act both as a mirror and a ground electrode. The other fibre optic plate is fixed

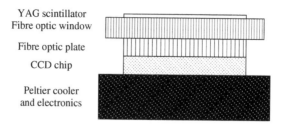

YAG scintillator
Fibre optic window

Fibre optic plate

CCD chip

Peltier cooler
and electronics

Figure 6.12 Schematic of an optically coupled CCD detector. (© Alan Craven.)

above the CCD. The fibre plate bonded to the scintillator often forms the vacuum window. The fibre plate attached to the CCD is optically coupled to the one forming the window with a suitable oil. This allows the CCD and its fibre plate to be removed when necessary. With this arrangement, only ten to twenty photoelectrons are created per incident electron giving a much larger dynamic range and a respectable DQE.

Although the uniformity of the device and the single read-out amplifier lead to good performance, it is worth looking at some practical limitations. In section 6.8.1, the dynamic range was defined as the ratio of the cell capacity to the read-out noise. In a CCD, the read-out noise is independent of the number of carriers in the cell. However, if the signal is equal to the read-out noise level, the DQE is very low. Thus the cell has to be filled to a much higher level if the signal to noise ratio is to be determined essentially by the shot noise in the incident signal. Thus matching the integration time to the input signal level is essential. However, if the integration time is such that some cells try to fill beyond their capacity, the excess carriers can spill into adjacent cells giving blooming unless the CCD has anti-blooming features to take such carriers to ground.

The scintillator/fibre optic/CCD arrangement allows some of the light created above one cell to reach neighbouring cells giving a point spread function (PSF). Careful coupling has reduced this in recent years but the point spread function is still present and still allows signal to leak into neighbouring cells albeit at much lower levels.

The cells have a dark current, i.e. an output signal that is present in the absence of an input signal. This is minimised by cooling the array. Commonly the CCD operates in the range $-10°C$ to $-20°C$ range and is cooled using a Peltier cooler but other arrangements are possible. Because of the nature of the semiconductor fabrication, the individual cells are very similar to each other but not quite identical. Each has a slightly different dark current. Thus it is necessary to record the dark current in the absence of the signal under the same conditions that the signal itself is recorded. Such a dark current reference is stored and can be subtracted from the signal. Of course only the average part of the dark current is removed while any noise present on the dark current reference is added to the noise in the signal. This is still an area that limits performance in applications such as spectrometry. One advantage of lens coupling demonstrated by Tencé is that the CCD camera can be cooled to a lower temperature, giving much reduced dark current and hence lower readout noise.

The dark current can change with time for a number of reasons. One is a change of the temperature of the CCD. Another is the result of the scintillator having long lived excited states in addition to those that decay rapidly, as noted in section 6.8.3. In regions where the scintillator is exposed to a high fluence, light continues to be emitted over minutes or even hours giving a contribution to the dark current which decays with time. This can be particularly important in electron energy loss spectrometry (see Chapter 7, section 7.2.2). Thus it is best to avoid exposing the scintillator to local high intensity for significant periods of time. Because of such effects, it is essential to renew the dark current reference frequently during the working day.

While the response of the cells is reasonably uniform, there is some cell to cell variation in sensitivity. In addition, the sensitivity of the scintillator can vary with position and, when fibre optic plates are used for coupling, there can be a complex Moiré pattern of light transmission. Such effects give rise to 'fixed pattern noise' in the sense that it not time dependent. This fixed pattern noise is stable and thus a gain reference can be created by uniformly illuminating the array. If a real image is divided by the gain reference, the fixed pattern noise is taken out to good accuracy. However, the fixed pattern noise will change if there is any relative motion between the different components of the detector. Changes will clearly occur if the detector is disassembled and re-assembled but can also occur if the detector is cycled to room temperature. After such changes, the gain reference should be re-acquired (section 7.2.2).

6.9 IMAGING USING TRANSMITTED ELECTRONS

6.9.1 The Diffraction Pattern

As mentioned in Chapter 3, section 3.4, one of the great strengths of STEM is that any signal can be used to image the specimen provided that the signal can be detected. As well as the transmitted electrons, backscattered electrons, secondary electrons, and photons are all examples of signals that can be used. Here, we concentrate on the transmitted electrons. Use of X-rays and inelastically scattered electron is considered in Chapter 7.

Emerging from the specimen is the diffraction pattern. This extends out to a cut-off imposed by some limiting aperture in the column. Figure 6.13a shows an example from an FEI Tecnai where the maximum angle of scattering reaching the diffraction plane is ∼230 mrad. A higher order Laue zone can be seen as a circle. The inner portion of the pattern corresponds to zeroth order Laue zone and the detail of this is shown in Figure 6.13b. If the crystal were perfect and at 0K, there would be no intensity outside the Bragg reflections. However, there are many mechanisms which transfer intensity into the regions between the Bragg reflections in a real specimen at finite temperature including deviations from crystalline perfection, thermal diffuse scattering from phonons and inelastic scattering from excitation of atomic electrons. Such inelastic scattering itself can be Bragg diffracted giving rise to the Kikuchi pattern. Note that the straight Kikuchi lines demonstrate that

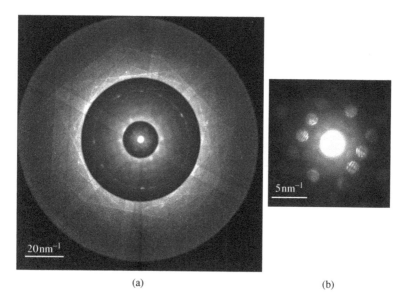

(a)   (b)

Figure 6.13 (a) A diffraction pattern showing that the maximum angle of scattering reaching the diffraction plane in an FEI F20 is ∼230 mrad. Relative to the inner region, the intensity of the mid region has been scaled by 8x and the outer by 80x (b) a shorter exposure showing the centre of the pattern. (Courtesy of S. McFadzean unpublished.)

the radial and spiral distortions in the post-specimen lenses are well compensated in the Tecnai column.

The centre of a diffracted disc is displaced from the centre of the incident disc by $2\vartheta_B = \lambda/a$ where ϑ_B is the Bragg angle and a is the corresponding plane spacing in the crystal. The angular diameter of the disc itself is 2α. Equation 6.1 gives the diameter of the Airy disc as $d_d = 1.2\lambda/\alpha$. Thus, if the discs in the diffraction pattern just touch, $d_d = 2.4a$. The implication is that the Airy disc is irradiating more than two repeats of the crystal periodicity. The simple criterion for resolution is that the spacing must be greater than half the diameter of the Airy disc. Thus, under these conditions, it might be expected that the crystal periodicity would not be resolved in the image (see section 5.3). If the discs in the diffraction pattern do not overlap, as in Figure 6.3e or i, the Airy disc is even bigger and so the periodicity will definitely not be resolved. The next section considers what happens when the discs overlap.

6.9.2 Coherent Effects in the Diffraction Pattern

It is worth thinking about this in terms the reciprocity theorem first (Chapter 3 section 3.3). In the CTEM, the smallest objective aperture that allows a periodicity to be resolved has a diameter slightly greater than $2\vartheta_B$. To achieve this, the illumination must be inclined at ϑ_B to the axis so that the incident beam falls close to the edge of the aperture and the diffracted beam appears close to the edge at the opposite side, i.e. the tilted dark field mode. In STEM, using an objective aperture just bigger than $2\vartheta_B$ causes the central and diffracted discs to overlap. Putting a detector at the centre of this overlap region corresponds to tilting the illumination by ϑ_B in the CTEM. Under these circumstances, the STEM image should give the same phase contrast image as the CTEM and so resolve the periodicity (see section 5.3.1).

In order to achieve this, the waves arriving at the overlap region in the diffraction pattern must be coherent, i.e. they must have a well defined phase relationship to each other. This happens if the Gaussian image of the source is much smaller than the Airy disc and hence is much smaller that the periodicity in the crystal. Formally, the coherence of the wave across the objective aperture is controlled by the Fourier transform of the intensity distribution in the Gaussian image. The Fourier transform of the Airy disc is just the aperture. If the Gaussian image is much smaller than the Airy disc, its Fourier transform is much wider than the aperture, which is then coherently illuminated (Barnett, 1974).

As shown schematically in Figures 6.14a and b, the diffraction process can be thought of as generating two coherent point sources of electrons whose separation decreases as the focus of the probe approaches the specimen. These two point sources give rise to Young's fringes in the overlap region of the diffraction discs with the fringe spacing increasing as the focus of the probe moves towards the specimen. To demonstrate this diffraction patterns were recorded as a function of defocus from Si oriented at the [110] pole. Figure 6.14c is the pattern averaged over

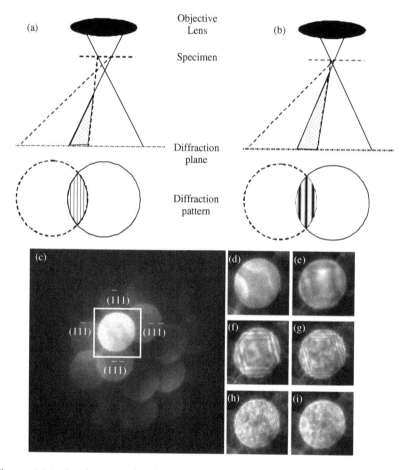

Figure 6.14 Production of coherent sources by diffraction: (a) schematic at a significant defocus with the fringe pattern in the overlapping discs below; (b) schematic at a smaller defocus with the fringe pattern in the overlapping discs below; (c) an experimental diffraction pattern from a small probe averaged over defocus; (d)-(i) section of the pattern starting at zero defocus and incrementing in steps of 100 nm. (Courtesy of B. Schaffer unpublished work.)

defocus showing four {111} discs overlapping the central disc. Here $2\vartheta_B$ is 12 mrad while 2α is 16 mrad. Figure 6.14d shows the overlap regions with the probe focused on the specimen. Ideally, this should give rise to uniform intensity in the overlap regions but dynamical diffraction effects introduce additional contrast. Figures 6.14e to i show that increasing the defocus in steps of 100 nm results in fringes whose spacing is inversely proportional to the defocus, as expected.

The positions of the fringes in the overlap region depend on the position of the probe on the specimen. The fringes displace by one fringe spacing as the probe moves across the specimen by one repeat distance. By placing a detector whose diameter is smaller than the fringe spacing in the overlap region, the intensity detected will oscillate as the probe is scanned. Thus the repeat distance is resolved in the STEM image, in agreement with the prediction of the reciprocity theorem (see also section 5.3.2 and Figure 5.3).

This was explained more directly by Zeitler and Thomson (1970). Figure 6.15a and b demonstrate the effect. Figure 6.15a shows the situation when the focused probe is positioned on an atom. Some of the amplitude in the ray arriving from the top left goes on in the forward direction while some of the amplitude arriving in the ray from the top right is Bragg scattered into the same direction (dashed line) allowing two coherent contributions interfere. Because the probe is positioned on the atom, there is no phase change due to a difference in the geometric paths of the rays. In Figure 6.15b, the probe has moved half a period. Wave fronts corresponding to the rays are shown as grey dotted lines normal to the rays. The particular wave fronts shown are those that reach one particular atom. It is clear the wave from the top left reaches the atom after travelling a different distance to that from the ray from the top right. This path difference is AO plus OB. Both segments have a length $\frac{1}{2}a\vartheta_B$ giving a total path difference of $a\vartheta_B$, which is just $\lambda/2$. This introduces an extra phase shift of π between the two contributions reaching the detector and this shifts the fringe system by half a fringe.

The rest of Figure 6.15 demonstrates this in practice. Figures 6.15c is an in-focus HAADF image while Figure 6.15d is a simultaneously recorded bright field image recorded with a circular detector whose angular diameter is similar to that of the diameter of the diffraction disc. The probe was then scanned along the line marked on the images and, at each pixel, a complete diffraction pattern was recorded. Figure 6.15e is the sum of all the patterns so that any interference effects are averaged out. Again $2\vartheta_B$ is 12 mrad while 2α is 16 mrad. Figures 6.15f to j show the overlap region on the central disc as the probe is moved in quarter

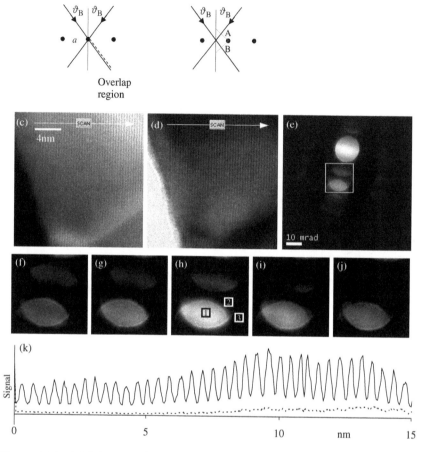

Figure 6.15 (a) Schematic diagram showing the focused probe positioned on an atom so that a forward scattered ray and a Bragg diffracted ray (dashed) travel to the overlap region in the diffraction pattern to interfere; (b) the situation when the probe is positioned half way between the atoms with the grey dotted lines representing the wave fronts; (c) HAADF image; (d) BF image: (e) average diffraction pattern from the pixels along the scan line; (f) – (j) enlarged regions of the pattern showing the variation of the interference of the overlapping coherent diffraction discs as the probe is moved in quarter unit cell steps; (k) profiles of the intensity in the overlapping disc area (Box 1 on (h)) and the non-overlapping disc area (Box 2) relative to the background (Box 3) confirming the variation in interference in the overlap region as the probe is scanned and its absence in the non-overlap region. (c)–(k). (Courtesy of B. Schaffer unpublished work.)

period steps. Since the probe is in focus, effectively only one point on the fringe system is seen and the intensity in the overlap region is constant. However, it is clear that this intensity changes as the probe moves, going through one cycle as the probe moves one period. Image line profiles can be obtained from the full dataset. Figure 6.15k shows two such line profiles. The solid trace is from the signal from the region corresponding to Box 1 on Figure 6.15h minus that from Box 3. Box 3 just provides a background signal from a region where there are no diffracted discs. It is clear that the specimen periodicity is imaged with high contrast. On the other hand, the signal from Box 2 minus Box 3, representing the non-overlapped region of the central disc, shows no sign of the periodicity.

One of the key benefits of aberration correction is that the probe has its minimum size very close to Gaussian focus since it is not necessary to compensate the effect of spherical aberration with defocus. Thus the periodicity of the interference fringes is always very large allowing the use of large bright field detectors without loss of contrast (Pennycook et al., 2009) as shown in Figure 6.15d, where the size of the bright field detector is approximately the same as that of the diffraction discs.

Returning to the reciprocity theorem, a CTEM can image a period using axial illumination when the objective aperture has an angular diameter slightly greater than $4\vartheta_B$. In the STEM, this corresponds to the outer edge of a diffracted disc overlapping with the centre of the central disc. This automatically means that there will be overlap from the disc arising from the opposite Bragg reflection giving further possibilities for interference. In the CTEM, the objective aperture size can continue to be increased provided there is sufficient spatial coherence (sufficiently parallel illumination), sufficient temporal coherence (sufficiently low energy spread) and sufficiently low image deflection from stray fields, vibration and drift. The same is true for the STEM. For each new set of diffraction spots allowed to pass through the aperture of the CTEM, a new set of coherent disc overlaps occurs in the STEM diffraction pattern. In this situation, the structure of the diffraction pattern becomes very complex, as seen in Figures 6.3b and f. Even if the aberrations were fully corrected, the pattern would change in a complex way with changes in the probe position, the defocus, the orientation and the thickness.

STEM allows detection of any part of the diffraction pattern emerging from the specimen. The signal level and the contrast obtained in the image depend on the angles of scattering collected by the detector. By the reciprocity theorem, the angular range of the detector in STEM corresponds to the angular range of incoherent illumination in the

CTEM. However, while it is straightforward to collect a large angular range in the STEM, it is much harder to illuminate with a large angular range in the CTEM even when using scanning of the illumination direction to increase the available angular range.

6.9.3 Small Angular Range – Bright Field and Tilted Dark Field Images

As noted above, a small detector placed on the axis gives a STEM image that is the same as the corresponding bright field CTEM image (Chapter 5, section 5.3.1). If the detector is placed off-axis, the STEM image gives a tilted dark field image. Early in the development of STEM, phase contrast and diffraction images were demonstrated (e.g. Colliex *et al.*, 1977). In an uncorrected CTEM, the angular range of the illumination has to be small to give coherent illumination. This implies that the STEM detector has to be small and thus the signal available is limited. The advent of aberration correction has allowed an increase in the angular range of illumination that can be used in the CTEM and this translates into STEM as allowing a bigger detector and hence higher signal, as noted the previous section. Hence there has been a resurgence of interest in the bright field image in the STEM (Pennycook *et al.*, 2009). This has been strengthened by the demonstration by Okunishi *et al.* (2009) that use of an annular detector in the bright field cone allows imaging of columns of light elements.

6.9.4 Medium Angular Range – MAADF

However, STEM is not restricted to the use of small detectors. By placing an annular detector on the diffraction pattern so that the bright field cone passes through the hole in the centre, another type of dark field signal can be obtained. The contrast in the signal depends on the sizes of the inner and outer angles of the detector. If the detector collects the scattering in the region corresponding to the zeroth Laue zone of a crystal (typically out to 25–50 mrad at 100 keV), it gives essentially the inverse signal to that from a detector collecting the whole of the bright field cone, i.e. one that corresponds to a diffraction contrast image taken with very convergent illumination in the CTEM. An annular detector collecting over this angular range is known as a medium angle annular dark field detector (MAADF) and gives information about boundaries, phases and defects. When the probe angle is small, it is very clear where to set the inner and outer angles of the detector. As the probe angle increases, the inner angle must increase to exclude the central disc and maintain the dark field imaging. The outer angle must also be increased by the same absolute angle to accept the full diffraction discs at the edge of the pattern. This means the ratio of the ideal outer and inner angles changes with probe angle whereas the actual ratio is normally fixed by the physical shape of the detector. Thus the detector must be fabricated to give the best compromise for the range of probe angles likely to be used.

6.9.5 High Angular Range – HAADF

As we have seen in the previous chapter, at large scattering angles, the atomic scattering factor from an atom tends towards that for Rutherford scattering from the nucleus. In addition, the phase shifts introduced by small random changes in the atom positions due to thermal vibration become large so that the coherent interference of the elastic scattering from the ideal crystal is lost. The detailed theory has been covered in the previous chapter and, to a first approximation, the signal is the incoherent sum of the intensities of the Rutherford scattering from all the atoms irradiated by the beam. Thus, provided the inner angle of the detector exceeds a critical value, the image intensity is proportional to the mean value of Z^ζ where Z is the atomic number and ζ is somewhat less than 2 by an amount depending on the actual value of the inner angle, (See Chapter 1, section 1.3). Such a detector is known as a high angle annular dark field (HAADF) detector and is a very powerful means of getting atomic resolution as shown in Figure 6.4.

In an uncorrected instrument operated at 100keV, α is $\sim$10 mrad and an inner angle of 60 mrad is sufficient to give good Z contrast. However, with a 5th order corrected instrument, values of α of 50 mrad are possible so that the inner angle has to be increased substantially to maintain good Z contrast. SuperSTEM2 (a Nion UltraSTEM) currently uses an inner angle of $\sim$100 mrad and an outer angle of $\sim$185 mrad.

6.9.6 Configured Detectors

So far, the signals from the detectors have been considered individually. However, the contrast from specific properties of the specimen can be enhanced by combining the signals from a number of regions in the diffraction pattern. Rose (1974) noted that phase contrast in a weak phase object reverses in sign as the angle of scattering increases. Thus it is possible to choose a series of annular detectors so that the sign of the phase contrast alternates from annulus to annulus. By adding all the signals together with the signals from every second annulus reversed, the total signal can be increased significantly.

Configured detectors giving differential phase contrast imaging have played the biggest role. First proposed by Dekkers and de Lang (1974), such a detector uses the fact that the phase from a weak phase object changes sign across the axis, i.e. the forward scattering direction. Thus if an axial detector is split in half and the two signals are subtracted, the signal is proportional to the phase gradient which can be mapped as a function of position. This technique has found its major application in Lorentz microscopy where the component of the magnetic induction perpendicular to the electron probe introduces a phase gradient across the probe which manifests itself as a small angular deflection of the probe. Thus one side of the detector receives more of the central disc of the diffraction pattern than the other. The difference is proportional to the magnitude of the induction and its direction relative to the split. By further splitting the detector into quadrants rather than halves, the in-plane component of the field can be mapped in both magnitude and direction. Since the information on the magnetic field comes from the edge of the detector, removing the central part to give an annular quadrant detector increases the contrast and reduces the noise. The phase of the electrons is also modified by physical structure of the thin film. An eight segment detector composed of a quadrant detector surrounded by an annular quadrant detector helps to separate the magnetic and structure contributions (Chapman et al., 1992). A better way to isolate the magnetic structure is to record the whole diffraction pattern at each position and determine the position of the central disc irrespective of the variation of intensity within it. The rapid developments in camera technology, image processing and storage look as though this will become a practical approach in the near future.

However, to see the magnetic structure, the specimen has to be in field free space or in a controlled low magnetic field. This is not the case in a typical high resolution electron lens where the magnetic field at the specimen is of the order of a Tesla. When a lens produces the probe outside its lens field, the $C_{3,0}$ is large and aberration correction may offer substantial benefits here.

6.10 SIGNAL ACQUISITION

It is clear that there is a wide range of information available simultaneously in the diffraction pattern. For examples, the HAADF, MAADF, and BF signals are all available simultaneously. In principle, detectors can be stacked one after the other as shown schematically in Figure 6.16a and detect the regions of the diffraction pattern shown in Figure 6.16b.

In practice, the physical construction of the detectors limits what can be achieved. With SuperSTEM2, the HAADF detector and either the MAADF or the BF detector can be used simultaneously. Figure 6.17a and b are high resolution HAADF and BF images recorded simultaneously while Figures 6.17c and d are MAADF and HAADF images

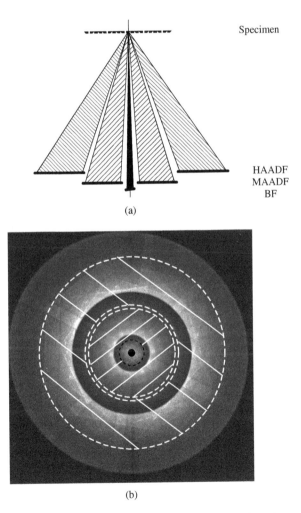

Figure 6.16   (a) Schematic stacking of HAADF, MAADF and BF detectors showing that all three signals are available simultaneously; (b) corresponding regions detected on the diffraction pattern. (© Alan Craven.)

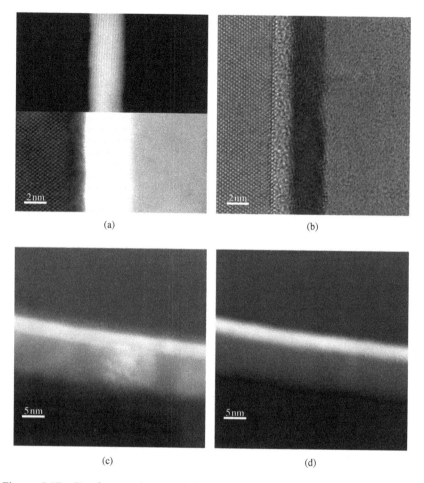

Figure 6.17 Simultaneously recorded images: (a) high angle annular dark field (HAADF); and (b) bright field (BF) images. The contrast is changed half way down the HAADF image to show detail in the HfO_2 in the upper part and Si in the lower part. Simultaneously recorded images: (c) medium angle annular dark field (MAADF); and (d) high angle annular dark field (HAADF). (Courtesy of B. Schaffer unpublished.)

recorded at lower magnification. The MAADF image shows significant diffraction contrast while it is almost totally suppressed in the HAADF image. Since the MAADF is available simultaneously with both the HAADF detector and the EELS signal, it is very useful for drift correction because of the relatively high contrast and signal level.

Cowley and Spence (1979) proposed that many detectors could be placed in the diffraction pattern to extract the signals that gave the most important information from a given specimen. They implemented an optical system for collection of many signals simultaneously so that bright field, tilted dark field, weak beam dark field, annular dark field images could in principle be recorded in parallel along with images making use of the diffuse scattering between Bragg spots. A similar approach can be made using a CCD camera, as illustrated in Figure 6.15. At each point on the specimen, a diffraction pattern can be recorded giving a multi-dimensional data set $I(x, y, k_x, k_y)$. Post-acquisition, each diffraction pattern can be masked to create images from the scattering angle ranges of interest or the position of a specific disc can be identified to give a direct measure of its deflection. In addition, the diffraction patterns from individual pixels or groups of pixels can be analysed in detail. However, the resulting datasets are extremely large and, with current CCD cameras, the read-out time involved makes the data acquisition a slow process. As faster array detectors become available, this will be an attractive way forward. In practice there will be limitations from limited dynamic range (e.g. difficulty in recording high contrast patterns), the point spread function (e.g. leakage of high intensity signal into regions of low intensity) and the background in the pattern from electrons multiply scattered in the system (e.g. setting a lower limit to sensitivity).

As well as offering multiple signals simultaneously, the diffraction pattern from a coherent probe giving rise to overlapping diffraction discs offers a resolution better than the normal limit imposed by the size of the Airy disc. This super resolution approach was pioneered by Rodenburg (2008).

Jeanguillaume and Colliex (1989) realised that in a system where the probe is scanned by the computer and spectra are acquired by the computer, the system can be set up to acquire a complete spectrum at each pixel in a technique that is termed spectrum imaging (SI). The spectrum imaging concept has now been extended to encompass recording a wide range of data at each pixel. Figure 6.15 showed an example where a diffraction pattern was recorded at each point. Systems are available to record the EELS and EDX spectra along with the signal from the HAADF image detector (Chapter 7), all of which are available simultaneously. Again the time to record such a dataset is relatively long. Incorporating drift correction limits the distortion of the image and the simultaneously acquired image ensures perfect registration between the spectral data and position is obtained.

Both the low loss and the core loss regions of the EELS spectrum should be recorded to allow full processing of the data. To ensure perfect registration, both should be recorded at the same pixel before moving to the next pixel. However, the large dynamic range of the EELS signal means that a single integration time cannot be used to record the whole spectrum. With current CCD camera controllers, changing the integration time incurs a large time overhead which becomes prohibitive in a spectrum image. Thus ways of coping with the dynamic range without changing the camera controller are being developed (e.g. Scott *et al.*, 2008). Figure 6.18 shows an example of a data set containing low-loss and core-loss EELS spectra, the EDX spectrum and the HAADF image.

Since damage to the specimen is caused by the incident electrons, the overall aim is to acquire the maximum information per incident electron. The discussion above shows that we are moving steadily toward this aim by acquiring signals in parallel where possible. However, this is not always possible. At present, there is no way of recording the diffraction pattern at the same time as an EELS spectrum. To do that would require an array of pulse height analysing detectors, each with the ability to determine the electron energy with an accuracy of parts per million. Unfortunately, nothing like this is on the horizon and the two must be recorded sequentially. What is developing is the ability to orchestrate the signal acquisition so that those signals of key interest in a particular experiment can be recorded with the necessary spatial registration in as efficient a way as possible. Here, it is important that manufacturers build in flexibility to allow detectors to be used in the optimum way for any particular problem. It is also essential to continue developing detectors with higher dynamic range and faster read-out time. The use of fast beam blanking to remove the probe from the area of interest when signals are not actually being recorded is also important.

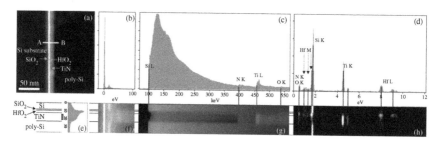

Figure 6.18   (a) HAADF survey image with the line AB showing the position of the line spectrum images (SI); (b) average low loss (LL) spectrum; (c) average background subtracted core loss (CL) spectrum; (d) average EDX spectrum; (e) HAADF signal recorded simultaneously with the SI and schematic of the specimen structure; (f) LL SI; (g) CL SI; EDX SI. (Reprinted from Scott *et al.*, Near-simultaneous dual energy range EELS spectrum imaging, Ultramicroscopy, Vol. 108, p 1586, Copyright 2008, with permission of Elsevier.)

Getting the most information per incident electron is not the same as getting the most information per unit recording time. STEM acquisition is point by point and thus takes a significant time. The increased probe current available with aberration correction reduces the time as do detector systems with faster read-out but the inherent sequential recording of the pixels will always require a 'long' acquisition with the requirement to select regions of interest carefully, having identified them by other means if possible, e.g. using some form of imaging or energy filtered TEM (EFTEM).

To illustrate this, consider an EELS spectrum image consisting of $10^3 \times 10^3$ spatial pixels and 10^3 energy steps. The same dataset can, in principle, be acquired using either STEM or EFTEM. In the latter case, all the spatial pixels in one energy slice are recorded in one exposure and the exposure repeated at each energy loss step. If we assume the datasets are recorded using sources with the same brightness, have the same incident and collection angles and contain the same intensity, then the STEM dataset will take 10^6 units of time and give 1 unit of dose to each spatial pixel while the EFTEM dataset will take only 10^3 units of time but will give 10^3 units of dose. There are practical differences here as well. For the highest spatial resolution, EFTEM has difficulty matching the spatial resolution (but this is changing with the advent of chromatic aberration correction), matching the illumination angles and maintaining the same current density available in the STEM. On the other hand, EFTEM is excellent for mapping unknown elemental distributions using a relatively low dose and so is good for finding regions suitable for more detailed analysis by STEM.

Thus, STEM will continue to develop as an immensely powerful technique for pico-analysis because it enables the maximum information per incident electron to be acquired with good spatial registration and allows the orchestration of the acquisition so that the regions from which the data are collected are tailored to the problem in hand.

Acknowledgements

I would like to thank the following for permission to include their unpublished work to illustrate key points in this chapter: Michael Finnie (Figure 6.2), Bernhard Schaffer (Figures 6.3, 6.8, 6.14, 6.15, 6.17), Mhairi Gass (Figure 6.4), Sam McFadzean (Figures 6.9, 6.13). I would also like to thank the members of the SuperSTEM team for constructive criticisms and suggestions and the many colleagues over the years who have contributed to my understanding.

REFERENCES

Barnett, M.E. (1974) Image Formation in Optical and Electron Transmission. *J. Microsc.*, 102, 1.

Chapman, J.N., Ploessl, R. and Donnet, D.M. (1992) Differential Phase Contrast Microscopy of Magnetic Materials. *Ultramicroscopy*, 47, 331.

Conrady, A.E. (1919) Star-Discs. *Monthly Notices of the Royal Astronomical Society* 575 June.

Colliex, C., Craven, A.J. and Wilson, C.J. (1977) Fresnel Fringes in STEM. *Ultramicroscopy*, 2, 327.

Cowley, J.M. and Spence, J.C.H. (1979) Innovative Imaging and Microdiffraction in STEM. *Ultramicroscopy*, 3, 433.

Craven, A.J. and Buggy, T.W. (1981) Design Considerations and Performance of an Analytical STEM. *Ultramicroscopy*, 7, 27.

Craven, A.J., Wilson, J. and Nicholson, W.A.P. (2002) A Fast Beam Switch for Controlling the Intensity in Electron Energy Loss Spectroscopy. *Ultramicroscopy*, 92, 165–180.

D'Alfonso, A.J., Cosgriff, E.C., Findlay, S.D., Behan, G., Kirkland, A.I., Nellist, P.D. and Allen, L.J. (2008) Three-Dimensional Imaging in Double Aberration-Corrected Scanning Confocal Electron Microscopy, Part II: Inelastic Scattering. *Ultramicroscopy*, 108, 1567.

Dekkers, N.H. and de Lang, H. (1974) Differential Phase Contrast in a STEM. *Optik*, 41, 452.

Gai, P.L., and Boyes, E.D. (2009) Novel In Situ Aberration Corrected Electron Microscopy at 0.1 nm Resolution. *Microscopy Research and Technique* 72: 153.

Jenguillaume, C. and Colliex, C. (1989) Spectrum-Image: The Next Step in EELS Digital Acquisition and Processing. *Ultramicrsocopy*, 28, 252.

Krivanek, O.L., Nellist, P.D., Dellby, N., Murfitt, M.F., and Szilagyi, Z.S. (2003) Towards Sub-0.5 A Electron Beams. *Ultramicrsocopy*, 96, 229.

Krivanek, O.L., Dellby, N., Keyse, R.J., Murfitt, M.F., Own, C.S. and Szilagyi, Z.S. (2008a) Chapter 3: Advances in Aberration-Corrected Scanning Transmission Electron Microscopy and Electron Energy-Loss Spectroscopy in: *Advances in Imaging and Electron Physics*, 153, 121 ed. Hawkes P.W. (New York: Academic Press).

Krivanek, O.L., Corbin, G.J., Dellby, N., Elston, B.F., Keyse, R.J., Murfitt, M.F., Own, C.S., Szilagyi, Z.S. and Woodruff, J.W. (2008b) An Electron Microscope For the Aberration-Corrected Era. *Ultramicroscopy*, 108, 179.

Muller, D.A., Kirkland, E.J., Thomas, M.G., Grazul, J.L. Fitting, L. and Weyland, M. (2006) Room Design for High-Performance Electron Microscopy. *Ultramicroscopy*, 106, 1033.

Okunishi E., Ishikawa, I., Sawada H., Hosokawa F., Hori M. and Kondo Y. (2009) Visualization of Light Elements at Ultrahigh Resolution by STEM Annular Bright Field Microscopy. *Microsc. Microanal.*, 15(2), 164.

Pennycook, S.J., Chisholm, M.F., Lupini, A.R., Varela, M., Borisevich, A.Y., Oxley, M.P., Luo, W.D., van Benthem, K., Oh, S-H., Sales, D.L., Molina, S.I., García-Barriocanal, J., Leon, C., Santamaría, J. Rashkeev, S.N. and Pantelides, S.T. (2009) Aberration-Corrected Scanning Transmission Electron Microscopy: From Atomic Imaging and Analysis to Solving Energy Problems. *Phil. Trans. R. Soc. A.*, 367, 3709.

Press, W.H., Teukolsky, S.A., Vetterling, W.T. and Flannery, B.P. (1992) *Numerical Recipes in C: The Art of Scientific Computing* (2nd edn), Cambridge University Press, Cambridge, Ch 12.1, p 500.

Rodenburg, J.M. (2008) Ptychography and Related Diffractive Imaging Methods. *Advances in Imaging and Electron Optics*, 150, 87.

Rose, A. (1948) Television Pickup Tubes and the Problem of Vision. *Adv. in Electronics*, 1, 131 ed. Marton, L., New York: Academic Press.

Rose, H. (1974) Phase-Contrast in Scanning Transmission Electron Microscopy. *Optik*, 39, 416.

Sader, K., Schaffer, B., Vaughan, G., Brydson, R., Brown, A. and Bleloch, A. (2010) Smart Acquisition EELS. *Ultramicroscopy*, 110, 998.

Scott, J., Thomas, P.J., MacKenzie, M., McFadzean, S. Wilbrink, J., Craven A.J. and Nicholson, W.A.P. (2008) Near-Simultaneous Dual Energy Range EELS Spectrum Imaging. *Ultramicroscopy*, 108, 1586.

Van Benthem, K., Lupini, A.R., Oxley, M.P., Findlay, S.D. Allen, L.J. and Pennycook, S.J. (2006) Three-Dimensional ADF Imaging of Individual Atoms By Through-Focal Series Scanning Transmission Electron Microscopy. *Ultramicroscopy*, 106, 1062.

Zeitler, E. and Thomson, M.G.R. (1970) Scanning Transmission Electron Microscopy I & II. *Optik*, 31, 258.

7

Electron Energy Loss Spectrometry and Energy Dispersive X-ray Analysis

Rik Brydson and Nicole Hondow

Leeds Electron Microscopy and Spectroscopy (LEMAS) Centre, Institute for Materials Research, SPEME, University of Leeds, Leeds, UK

One of the chief benefits of a scanning configuration in the transmission electron microscope is the ability to use the highly focused probe to undertake elemental and/or chemical analysis of a particular region of the micro or nanostructure at an extremely high spatial resolution. In addition, this analytical information may be recorded as a function of position to form a one dimensional linescan, a two dimensional map or, in some cases, a three dimensional tomogram. As we will see in this chapter, aberration correction of the probe significantly enhances this analytical capability both in terms of spatial resolution and in terms of the detectability of a specific signal.

In the transmission configuration, elemental and/or chemical analysis using electrons is usually undertaken using either the direct analysis of the energies of the transmitted electrons or the analysis of the energies of the X-rays emitted as a result of the electron-sample interaction. These two techniques are the subject of this chapter.

Aberration-Corrected Analytical Transmission Electron Microscopy, First Edition.
Edited by Rik Brydson.
© 2011 John Wiley & Sons, Ltd. Published 2011 by John Wiley & Sons, Ltd.

7.1 WHAT IS EELS AND EDX?

As has been highlighted in Chapter 1, the high energy incident electron beam of the TEM interacts with the sample both elastically and inelastically. For elemental and chemical analysis we are concerned with inelastic processes such as: **Phonon scattering, Plasmon scattering, Single electron excitation and Direct Radiation Losses.** Electron energy loss spectrometry (EELS) is an *absorption spectroscopy* in which we directly measure the energy losses of the incident electrons following transmission through the thin sample. A complementary secondary emission process provides the basis for an *emission spectroscopy* which monitors the energy or wavelength of the emitted X-rays produced as a result of the interaction of the incident electron with the sample – known as energy dispersive or wavelength dispersive X-ray (EDX or WDX) analysis. In transmission electron microscopy we are principally concerned with EDX rather than WDX.

7.1.1 Basics of EDX

In a TEM sample, atoms which have undergone inner shell ionization by the primary electron beam and have been promoted into a higher energy excited state, relax back to their ground state by a process in which electrons from higher energy levels drop into the hole created in the vacant inner shell. As introduced in section 1.2.2.2, this relaxation process results in the release of excess energy corresponding to the difference between the electron energy levels involved in the transition (ΔE) and this occurs via the creation of either a low energy (100–1000 eV) *Auger* electron or, alternatively, an X-ray or visible photon of wavelength $\lambda = hc/\Delta E$ as shown in Figure 7.1. The Auger yield is generally small, except for the case of light elements, whilst the alternative de-excitation mechanism favoured by heavier elements involves the emission of an X-ray. The energy of the X-ray photon emitted when a single outer electron drops into the inner shell hole is given by the difference (ΔE) between the energies of the two excited states involved. This energy difference is characteristic of the atom and allows for elemental identification, whilst the X-ray intensity can be used for quantification of the elemental concentration. For example (see Figure 7.2), if a K-shell electron is ionised from a nickel atom and is replaced by a higher energy L-shell electron falling into the vacant state the energy difference, ΔE, is ca. 7500 eV which is emitted as a Kα X-ray of Ni; the X-ray wavelength, given by $\lambda = hc/\Delta E$, is 0.1658 nm. Alternatively, the inner K-shell hole

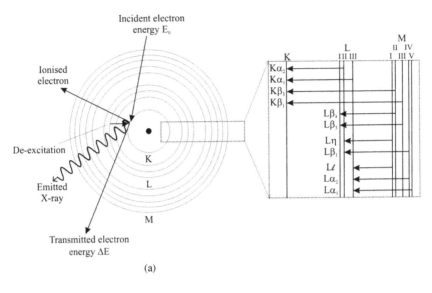

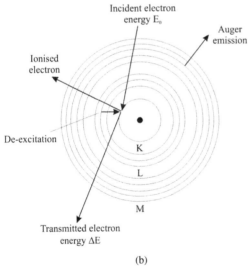

Figure 7.1 De-excitation mechanisms for an atom which has undergone K-shell ionization by primary electrons: (a) emission of a characteristic $K\alpha$ X-ray (inset details possible ionisation processes and emission of X-rays) and (b) emission of a KLM Auger electron.

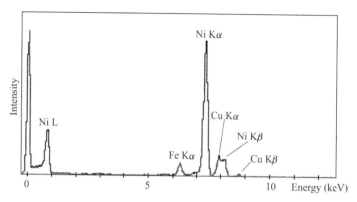

Figure 7.2　Schematic energy dispersive X-ray (EDX) emission spectrum from a sample of nickel nanowires with an iron contaminant on a carbon support film. The copper Kα and Kβ X-ray peaks are due to the TEM specimen holder and support grid. (Reprinted from A S Walton, *et al.*, Four-probe electrical transport measurements on individual metallic nanowires, *Nanotechnology*, Vol. 18, 065204, (6pp), Copyright 2007, with permission of IOP.)

could be filled by an electron dropping from the higher energy Ni M-shell which would result in the emission of a Ni Kβ X-ray of energy 8260 eV and wavelength 0.1500 nm. The exact nomenclature of X-ray emission spectroscopy involves the use of the term K, L, M, N, O etc. which denotes the inner (Bohr) shell which has undergone ionisation. There is a further subsequent term, α, β, γ ..., etc., as well as a subscript, 1, 2, etc., which denotes the family of lines formed by all the possible electron transitions (Figure 7.1). At first sight, it would appear that there would be a large set of characteristic X-rays for each atom, since electron transitions between all possible energy levels would appear to be allowed. However, a set of dipole selection rules determine which transitions are observed; more specifically the change in angular momentum (quantum number) between the two electron states involved in transition must be equal to ±1. This prohibits many of the expected transitions, such as a transition between a 2s state and a 1s state. Numerous general electron microscopy texts such as Jones (1992) or Williams and Carter (2009) list the most useful X-rays for microanalysis in the TEM.

7.1.2　Basics of EELS

Apart from the emitted X-ray signal from a TEM sample, the inelastically scattered component of the transmitted electron beam directly reflects the response of the electrons in the sample to the electromagnetic disturbance introduced by the incident electrons. This inelastically scattered intensity may be spectroscopically dispersed as a function of electron kinetic energy and hence energy loss from the incident beam energy using an electron spectrometer and recorded in parallel using a scintillator and either a one or two dimensional photodiode array detection system.

The various energy losses observed in a typical EEL spectrum have been shown in Figure 1.3 (Chapter 1). Overall, one of the most striking features is the large dynamic range of the spectrum – the intensity above 1000 eV energy loss is typically seven or eight orders of magnitude less than the intensity of the peak at the zero energy loss. The latter is known as the *zero loss peak* which, in a specimen of thickness less than the mean free path for inelastic scattering (roughly 100 nm at 100 keV), is by far the most intense feature in the spectrum and contains all the elastically and quasi-elastically (i.e. vibrational- or phonon-) scattered electron components. The full width half-maximum (FHWM) of the zero loss peak is taken as a measure of spectral resolution and, neglecting the effect of the spectrometer and detection system, this is usually limited by the energy spread inherent in the electron source which, following section 2.5, can vary between ∼0.3–2 eV, depending on the type of emitter. In recent years the use of electron monochromators has improved achievable resolutions to ∼0.1 eV.

The *low loss* region of the EEL spectrum is usually defined as extending from the zero loss peak to about 50 eV energy loss and corresponds to the collective excitation of electrons in the outermost atomic orbitals which, in a solid, are delocalized due to interatomic bonding and may extend over several atomic sites. This region therefore reflects the solid state character of the sample. The smallest energy losses (10–100 meV) arise from phonon excitation, and, as discussed in Chapter 1, these are usually subsumed in the zero loss peak. The dominant features in the low loss spectrum arise from collective, resonant *plasmon* oscillations of the valence electrons and (interband) transitions from valence to conduction bands.

The *high loss* region of the EEL spectrum extends from about 50 eV to several thousand electron volts and corresponds to the excitation of electrons from localised orbitals on a single atomic site to extended, unoccupied electron energy levels just above the Fermi level of the material. This region therefore more reflects the atomic character of the specimen and exhibits ionisation edges, corresponding to the excitation of inner shell electrons, superimposed on a monotonically decreasing background intensity principally due to both collective and low energy single electron excitations. The various ionisation edges are classified using the standard spectroscopic notation similar to that employed for labelling X-ray emission peaks; e.g., K excitation for ionisation of 1s electrons, L_1 for 2s, L_2 for $2p_{1/2}$, L_3 for $2p_{3/2}$ and M_1 for 3s etc.[1] The detailed *fine structure* associated with a particular ionisation edge is determined by the degree of quantum mechanical overlap between the initial- and final-state electronic wavefunctions involved in the excitation process and also the density of empty electronic states available to accept the ionised electron which changes as a function of chemical bonding.

7.1.3 Common Features For Analytical Spectrometries

Here we briefly highlight a number of common aspects of the two techniques which are especially important if we are to compare these to other analytical methods. Firstly, the *spectral resolution* governs the ease by which two spectral signals can be separated and independently analysed. We shall see that the spectral resolution in EELS is significantly better than that in EDX typically by a factor of 100 or more. However, both the spectral resolution and the *spectral range* govern the nature of the energy processes which can be easily measured and in EDX we have a considerably greater spectral range (some ten to twenty times larger) than in EELS which significantly compensates for the poor spectral resolution.

In very general terms, in analytical transmission EM we are interested in measuring a certain *concentration* of a particular element (C_{atom} measured in atoms/nm^3) within a certain *thickness* (t in nm) of a thin specimen. These two quantities multiplied together represent the *areal density* ($C_{atom} \cdot t$ in atoms/nm^2), which is what is actually measured when the analysis is projected through the specimen thickness.

The *spatial resolution* of the analysis is determined by the *probe diameter* (d in nm and commonly taken as the full width tenth maximum which contains 90% of the integrated intensity within the probe) and *analysed volume* which involves consideration of both the incident probe diameter and the probe (beam) broadening during passage through the specimen. Elastic scattering is the main cause of this beam broadening which increases with average atomic number and thickness (typical sample thicknesses and incident beam energies result in beam broadenings of a few nm) as discussed in Chapter 1, section 1.5. Beam broadening can

---

[1] Note: the subscript, in for example $2p_{1/2}$, refers to the total angular momentum quantum number, j, of the electron which is equal to the orbital angular momentum, l, plus the spin quantum number, s. For a 2p electron, $l = 1$, and this can couple to the electron spin in one of two ways, i.e. $j = l + s = 1 + 1/2 = 3/2$ (L_3) or $j = 1 - 1/2 = 1/2$ (L_2).

be less in a crystalline specimen when it is oriented along a low index crystallographic zone axis, owing to the phenomenon of channelling down the atom columns. In the case of EDX, we collect all X-rays produced isotropically within the beam broadened volume, whilst for EELS in transmission where the signal is highly forward-peaked, the effective broadening and hence analytical volume can be reduced by using an angle limiting collector aperture. This can significantly improve the ultimate spatial resolution of EELS compared to that obtained with EDX.

The areal density of a particular element (in atoms/nm^2) multiplied by both the *(partial) cross-section* (σ which represents the probability an analytical signal will be generated for the given experimental conditions such as accelerating voltage and is usually expressed in nm^2/atom) and the total *probe current* (I_{probe} in nA) gives the *emitted signal* ($S_{emitted}$) in detected electrons or X-rays per unit time. The actual *measured signal* is the emitted signal multiplied by the *detection (quantum) efficiency (DQE)*.

This description is in fact actually not quite so simple since, in reality, we have beam broadening and we may need to consider the intensity distribution of the probe at different depths in the analysed volume.

One collects the measured signal for a given *acquisition time* (τ) and, assuming random signal production events, the noise level, N, is simply the square root of the number of signals counted. This 'shot noise' is combined with any noise due to the detection system itself which gives an overall signal to noise ratio (SNR). The signal we are interested in for analysis is invariably superimposed on some form of background which is usually another spectroscopic process. This background affects the *detectability* of the measured signal and needs to be removed prior to quantitative analysis.

In terms of detectability, there are two important quantities: firstly, the *minimum mass fraction* (MMF) which is the smallest composition (expressed in either weight% or atomic%) which is detectable and is defined by:

$$MMF = \frac{1}{\sqrt{\Delta S.(\Delta S/S_B).\tau}} \tag{7.1}$$

where ΔS is the (analytical) signal intensity above the background intensity, S_B ($\Delta S/S_B$ is then the signal to background ratio, SBR) and τ is the signal acquisition time. Generally for analytical transmission electron microscopy, the MMF is rather poor compared to many other analytical techniques principally due to either the low total signal which is detected (EDX) or the large background contribution (EELS).

Secondly, the *minimum detectable mass* (MDM) is the minimum number of atoms detectable in the analytical volume which is probed. The ability to form small, intense electron probes means that it is possible to analyse very small total sample masses and, for a given detectable MMF, this results in a low MDM which is one of the major benefits of analytical transmission electron microscopy; aberration correction can significantly improve this capability.

The overall analytical accuracy involves consideration of the accuracy of background removal, the accuracy of the calculation or determination of the (partial) cross-section as well as the signal to noise ratio in the spectrum.

7.2 ANALYTICAL SPECTROMETRIES IN THE ENVIRONMENT OF THE ELECTRON MICROSCOPE

7.2.1 Instrumentation for EDX

X-rays produced when the electron probe interacts with the specimen are most commonly detected from the incident surface using a low *take-off angle* EDX detector (i.e. the detector is roughly in the same plane as the sample, perhaps some 20 degrees to the horizontal – see Figure 7.3). This low angle configuration allows the detector to be brought close to the sample and also minimises the predominantly forward-peaked Bremstrahlung background contribution to the X-ray emission spectrum which is continuous in terms of energy (see section 1.2.2.2). Even though the detector is inserted to within a few mm of the sample surface, it collects only a very small proportion of the isotropically-emitted X-ray signal owing to its limited solid collection angle (typically the detector may present a solid angle of between 0.3 to 0.03 steradians – although high area silicon drift detector arrays are under development providing substantially higher solid angles). Often the specimen is tilted towards the detector (typically, in a CTEM, through a *tilt angle* of 15 degrees – see Figure 7.3) so as to provide a clear X-ray trajectory between the irradiated area and the detector, while the volume of the specimen which produces X-rays is controlled by the electron probe size (and hence the current in probe forming lens) as well as beam broadening within the specimen which increases with (amongst other things) thickness and average atomic number and decreases with microscope accelerating voltage. High take-off angle X-ray detectors

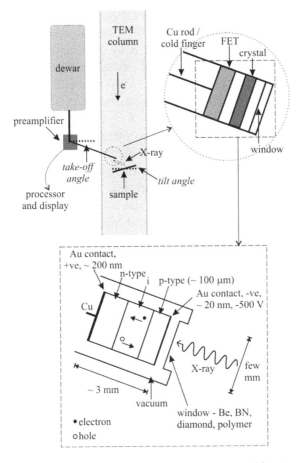

Figure 7.3   Schematic diagram showing the components and location of an EDX detector in a TEM.

also exist and these do not require any tilting of the specimen towards the detector.

Although EDX detectors should be positioned as close to the TEM specimen as possible so as to maximise the number of X-ray counts during the analysis of specific areas in a thin specimen, EDX detectors often need to be withdrawn or isolated via a shutter mechanism to prevent the detector being flooded with X-rays during the general illumination of larger and/or thicker specimen areas. Megahertz-rate beam blanking, the deflection of the beam off the specimen as soon as the EDX detects an incoming pulse (see section 6.7), is sometimes used to improve X-ray throughput (i.e. counts in versus counts out) of the detector system.

In addition to X-rays produced from the irradiated sample volume, it is important to realise that the X-ray detector can also detect additional X-ray signals produced as a result of stray electrons or X-rays striking microscope apertures (especially the objective aperture which needs to be removed prior to EDX analysis or in STEM a virtual objective aperture is often employed), objective lens polepieces, the (S)TEM sample holder and/or support grid as well as regions of the sample outside the primary region illuminated. Such signals can be partially estimated by collecting a *hole count* spectrum produced with the beam illuminating a hole in the specimen. Minimization of some of these effects can be achieved by using collimating condenser apertures, low specimen tilts during analysis as well as the use of low atomic number materials or coatings on polepieces, or for sample holders and support grids.

Ignoring the effects of stray and spurious scatter, owing to the well-defined nature of the various atomic energy levels, it is clear that the energies and associated wavelengths of the set of dipole-allowed X-rays that are emitted following electron-induced ionisation will all have characteristic values for each of the atomic species present in the irradiated specimen volume. Thus by measuring the energies of the X-rays emitted from the (top) surface of the sample, it is possible to determine which elements are present at the particular position of the electron probe and this is the basis for EDX analysis. EDX detectors collect X-rays in a near-parallel fashion and rely on the creation of electron-hole pairs in the central intrinsic region of a biased semiconductor *pn junction*; the number of electron-hole pairs and hence the current (which is amplified and converted to a voltage by a field effect transistor (FET)) is directly proportional to the energy of the incident X-ray (Figure 7.3). The detector has both a front and back electrode to collect the charge pulse and also a dead layer (i.e the p and n type region) on either side of the central intrinsic region. A reverse bias voltage to the p-i-n junction is applied through these thin gold electrodes deposited on the surfaces of the detector crystal – a 20 nm thick electrode provide sufficient conductivity whilst only absorbing a small number of incoming X-rays. Both the front electrode and the dead layer affect X-ray absorption and are thus important detector characteristics. The current generation of EDX detectors invariably employ ultra-thin windows in order to isolate them from the vacuum in the microscope column primarily to stop contaminants condensing on the cold detector crystal and therefore increasing the absorption of incoming X-rays. In addition to hydrocarbon contaminant layers, cooled detector windows can form layers of ice which periodically need to be removed by *conditioning* (heating up) of the detector. Thin polymer windows permit detection of elements as light as boron, although extensive absorption of low energy X-rays in the specimen may severely limit accurate quantification of first row elements in the Periodic Table.

A number of different detector crystals are currently employed: firstly, the most common are silicon doped lithium (Si(Li)) in which the lithium compensates for impurities in the silicon crystal; intrinsic high purity (hp) germanium detectors are also used which more strongly absorb high energy X-rays and thus give a larger X-ray energy range (40 or 80 keV in hpGe as opposed to 20 keV for Si(Li)). Detector crystals and the FET have to be cooled to low temperatures in order to reduce noise; furthermore in Si(Li) detectors cooling prevents diffusion of Li under the applied bias. Cooling is usually achieved using liquid nitrogen which, for Si(Li) crystals has to be constantly maintained in a dewar. In some cases cooling is achieved mechanically or thermoelectrically. Silicon drift detectors (SDD) are a recent development which employs an n-type Si crystal with rings of p-type material and anodes, this design gives a low capacitance which allows them to operate at relatively high temperatures ($-15°C$ which are easily reached thermoelectrically) and able to accept high count rates. Finally microcalorimeter detectors employ a completely different detection mechanism (they measure heat input arising from the X-rays) and these can provide very high spectral energy resolution, similar to wavelength dispersive X-ray analysers, which are capable of detecting small changes in X-ray energies arising from slight shifts in inner shell binding energies due to atoms in differing chemical environments.

Once electron-hole pairs have been created in the detector crystal, amplified and converted to a voltage by the FET, fast pulse-processor electronics allow separate voltage pulses to be discriminated and measured (and thus X-ray energies determined with a typical energy resolution of between ca. 120–140 eV for Si(Li) and hpGe detectors). As the count rate increases from a few thousand counts per second, an increasing number of pulses are rejected and the detector **dead time** increases. The important quantity for an X-ray spectrum is the **live time** not elapsed time. Typically a useful spectrum can be collected within about 1 minute. The discriminator levels and pulse processing speeds can be adjusted either to improve maximum count rates or alternatively improve peak separations and light element detection. Shorter process times improve spectral resolution at the expense of the overall count rate and normally the electronics are operated so as to optimise X-ray count rate.

7.2.2 EELS Instrumentation

By far the most common type of electron spectrometer which can be added to the end of either a STEM or CTEM column is the sector magnet spectrometer, shown in Figure 7.4. A typical sector magnet consists of a homogeneous magnetic field normal to the electron beam. This causes electrons of a given kinetic energy and hence velocity, to follow trajectories which are essentially arcs of circles. Usually for mechanical convenience the electrons are bent through a right angle and the magnetic force causes electrons of different energy (and therefore energy loss) to emerge from the spectrometer spatially dispersed. The spectrometer magnetic field is provided by two parallel pole pieces with two current carrying coils providing the excitation. In practice the pole pieces and coils are at atmospheric pressure, while an evacuated drift or flight tube, connected to the microscope vacuum system, passes between the pole pieces. An offset voltage may be applied to this drift tube which will alter the kinetic energy of all the electrons and rigidly shift the whole energy loss spectrum across the detector system. In fact the spectrometer, together with a set of manually adjustable quadrupole and sextupole lenses, has a double focusing action which results in the formation of a line spectrum (each energy loss is separately focused to a point) at the spectrometer image plane. The spectrometer dispersion (D in figure 7.4)

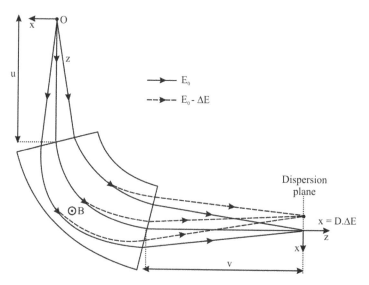

Figure 7.4 Schematic diagram showing the first order focusing properties and dispersion (D) for a magnetic sector EEL spectrometer.

is the displacement of the focus per unit change in electron energy and is typically a few micrometres per eV. The energy resolution of the resultant EEL spectrum is determined not only by the energy spread of the electron source, but also the resolution of the spectrometer itself, the effect of external interference as well as the resolution of the detection system. To minimise aberrations, the entrance angle of the spectrometer must be limited which is achieved by use of an adjustable spectrometer entrance aperture (SEA) at the position where the spectrometer bolts on to the CTEM camera chamber or, alternatively, by the use of an objective aperture (CTEM) or collector aperture (STEM) in the microscope to limit the collection angle of scattered electrons. The electron beam path must be also adequately screened from oscillating magnetic fields from both the mains supply and from high frequency oscillations in the microscope; besides enclosing the camera chamber and spectrometer in a soft magnetic material, this may be further achieved by adding an additional *ac* compensation signal into the spectrometer excitation coils.

The magnetic sector spectrometer can be experimentally coupled to a microscope in a number of ways. The simplest case is that of a dedicated STEM with the specimen in field-free space, as shown in Figure 7.5a. Here, the electron probe is focused onto the specimen with a semi-angle of convergence, α typically in the range 2–15 mrad for a non-aberration corrected STEM, but considerably higher for the case of an aberration corrected probe. In most dedicated STEMs, there are no lenses after the specimen and the specimen is in the object plane of the spectrometer. The area irradiated by the probe becomes the spectrometer object and its size is determined by the size of the electron probe, which can be as low as 0.1 nm in an aberration corrected instrument, and this effectively determines the spatial resolution for analysis. The collection semi-angle, β, of electrons entering the spectrometer is either defined by the spectrometer entrance aperture or alternatively, in a dedicated STEM, a post-specimen collector aperture may be employed. If the magnetic field due to the STEM objective lens is strong and immerses the specimen and there will be magnetic induction after the specimen and this will effectively act as a post-specimen lens, giving rise to the situation in Figure 7.5b; this is the same as in many hybrid CTEM/STEM instruments operated in STEM mode. Here the angular range (β) of electrons leaving the specimen can be compressed into a spectrometer of smaller angular acceptance by forming a suitably magnified (virtual) image of the specimen which acts as the spectrometer object; here β will be proportional to the radius of the spectrometer entrance aperture divided by the camera length of the diffraction pattern.

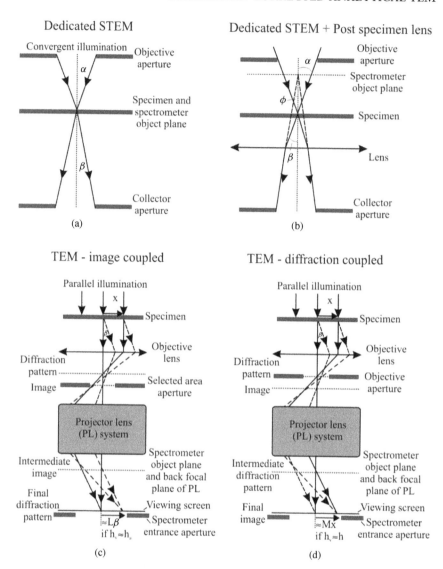

Figure 7.5 Schematic ray diagram of coupling between a magnetic prism spectrometer and: (a) a dedicated STEM; (b) a dedicated STEM with a post specimen lens; (c) a TEM-image coupled; and (d) a TEM -diffraction coupled. The diagram indicates the parameters which determine the area of analysis and the collection angle. (Reprinted from Rik Brydson, Electron Energy Loss Spectroscopy, BIOS Scientific Publishers, Oxford, Copyright 2001 with permission of Oxford University Press.)

Using EELS in CTEM mode is much less straightforward than for STEM, particularly for the case of analysis at high spatial resolution. In CTEM mode it is possible to operate the microscope in either diffraction mode or image mode. In CTEM diffraction mode, shown in Figure 7.5c, additional lenses after the specimen allow the angular compression (i.e. the camera length of the diffraction pattern) to be varied over a wide range. The spectrometer object is then the demagnified image of the specimen in the final projector lens and so the spectrometer is said to be *image coupled* (even though the CTEM is operating in diffraction mode). As for STEM with a post specimen lens, the collection semi-angle, β, is determined by the selected spectrometer entrance aperture and the diffraction camera length. The exact area from which the spectrum is taken is limited either by the area of the illuminated region or by the size of the selected area diffraction aperture (SAED, which can be as low as 200 nm in some instruments). However both spherical and, more importantly, chromatic aberrations in both the CTEM objective and projector lenses can affect the precision of the area selected using the SAED aperture and, for high spatial resolution analysis, limiting the area illuminated using a highly convergent probe (e.g. microprobe or nanoprobe mode) is generally the preferred option.

Operating the CTEM in image mode (so called *diffraction coupling*), illustrated in Figure 7.5d, is a less usual mode of operation (apart from in energy filtered TEM (EFTEM) imaging). Here the spectrometer object is the diffraction pattern at the crossover of the final projector lens. The collection semi-angle can normally only be defined accurately by inserting an objective aperture, although this may interfere with simultaneous EDX measurements owing to the stray X-ray signal generated from the aperture. The area of analysis on the specimen is usually controlled by the part of the image selected by the spectrometer entrance aperture; this depends inversely on the image magnification. However, if the irradiated area in the image is smaller than that selected by the spectrometer entrance aperture, this will ultimately control the analytical area but may cause problems due to chromatic aberrations affecting the collection efficiency of the system at different energy losses.

As stated, the spectrometer produces a dispersed EEL spectrum in its image (dispersion) plane and, in order to record the intensity at each particular energy loss as a sequence of digital values over a large number of detector channels (typically 1024 or 2048), most commonly a multi-element detector is employed where the whole spectrum is recorded in parallel. Detector systems have been discussed in detail in section 6.8. Modern one dimensional linear photodiode arrays (PDAs) and two dimensional arrays of charge coupled diodes (Ccds), the latter identical to those used for the digital recording of CTEM images, are both well suited to the task of parallel recording. As discussed in section 6.8.5, these systems rely on the measurable discharge (over a certain integration time) of a large array of self-scanning, cooled silicon diodes by photons created by the direct electron irradiation of a suitable scintillator prior to the CCD (Figure 6.12). This signal is superimposed on that due to thermal leakage currents as well as inherent electronic noise from each individual diode and together these are known as **dark current** which can be subtracted from the measured spectrum; **gain variations** and cross-talk (the **point spread function**) between individual diode elements can also be measured and corrected for. One major problem with such recording arrays is their limited dynamic range, which may require at least two or three suitable integration periods, and hence separate measurements, to record accurately and efficiently the entire EEL spectrum (under the same experimental conditions) which can show a total variation in signal from low loss to core loss of up to 10^8. This can cause problems in terms of assigning absolute energy losses to spectral features, however recent advances in detector technology have enabled the near simultaneous measurement of different spectral regions – so called 'dual EELS' (Gubbens et al., 2010). EELS data from the 1D or 2D detector array is collected, stored and then processed on a multichannel analyser software package.

7.2.3 Microscope Instrumentation for Analytical Spectroscopies

7.2.3.1 General Aspects

Generally for analytical spectroscopy at high spatial resolution, STEM is superior to CTEM for the following reasons:

- Firstly, the use of a scanned, focused probe rather than a broad parallel beam naturally lends itself to the generic technique of *spectrum imaging*, where a complete (or partial) EELS or EDX spectrum is recorded at each probe position as the beam is scanned. Spectral processing (often on-line) then allows the production of one-dimensional line profiles or two dimensional maps of a specific analytical signal.
- Secondly, digital control of beam scanning in STEM allows complex shapes or distributions within an image such as a second phase, a set of interfaces or surfaces to be analysed by EELS or EDX using so called 'smart acquisition' procedures – see section 6.4.5 (Sader et al., 2010).

- Thirdly, the STEM configuration is inherently enabled for multiple signal detection (section 3.4) – it is possible to do simultaneous imaging and EELS or EDX spectroscopy in STEM which is virtually impossible in CTEM. Whilst imaging the sample using the STEM (high angle) annular dark field (ADF) detector, the central bright field (BF) detector can be removed and the EEL spectrum recorded from the specific region being imaged. The same is true for EDX spectra and both BF and ADF STEM imaging. Furthermore, in a dedicated STEM, the use of a virtual objective aperture allows EDX spectra to be recorded whilst viewing BF images with enhanced diffraction contrast.
- Fourthly, for STEM, the coupling of the EELS spectrometer to the microscope is inherently simpler both in terms of the area analysed and the collection semi-angle which is accepted into the spectrometer.

7.2.3.2 Developments In Instrumentation

In terms of analytical spectroscopy, the principle benefit of current developments in aberration correction is solely in terms of probe correction, allowing the formation of smaller probes containing increased probe current (Figure 7.6). The smaller probe containing a given current density

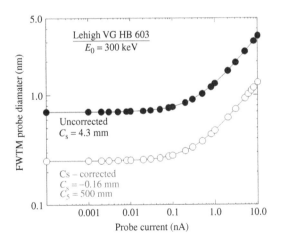

Figure 7.6 The simulated FWTM (full width tenth maximum) probe-size in the uncorrected (closed circle: $C_s = 4.3$ mm and $\alpha = 6$ mrad) and Cs-corrected (open circle: $C_s = -0.16$ mm, $C_5 = 500$ mm, and $\alpha = 15$ mrad) HB 603 (300 keV), plotted against the probe current. Note that the probe size is almost independent of the probe current up to 0.1 nA for both the uncorrected and Cs-corrected conditions. (Reprinted from M. Watanabe, et al., Improvements in the X-Ray Analytical Capabilities of a Scanning Transmission Electron Microscope by Spherical-Aberration Correction, *Microsc. Microanal.* Vol. 12, p. 515–526, Copyright 2006, with permission of Cambridge University Press.)

allows analytical signals, including both EDX and EELS, to be obtained from smaller volumes with improved (i.e. reduced) MDMs. The MMF may also be improved if the increased current density in a probe of a given size results in a shorter signal acquisition time for a given level of statistics – provided, of course, that there is no electron beam induced sample damage. Watanabe *et al.* (2006a) and Herzing *et al.* (2008) have estimated a reduction in the MDM of between 2–5 times (Figure 7.7) and an threefold improvement in the MMF (Figure 7.8) for probe corrected dedicated STEM/EDX at 300 kV relative to an uncorrected STEM, due to the tenfold increase in the probe current in the corrected machine. Single atom detection has been demonstrated using probe corrected STEM/EELS (see Figure 8.16), but, at the time of writing, not yet for STEM/EDX (where the detection limit is a few atoms and currently the spatial resolution is about 0.4 nm (Watanabe, 2009)). Note, more generally for EDX, the increased probe currents will lead to increased X-ray count rates which may necessitate the use of high throughput detectors such as SDDs or, alternatively, beam blanking can be used if the detector input signals for EDX or even low loss EELS are too high

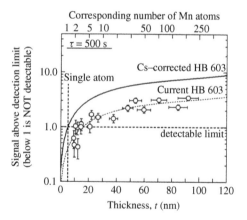

Figure 7.7 The detection limit of Mn atoms in a thin foil of Cu-0.12 wt% Mn solid solution, obtained in the uncorrected HB 603 indicated as closed circles. Under these conditions ∼2 Mn atoms are detectable in a 15-nm-thick region. The solid line indicates the detection limit calculated with the same probe size (1.1 nm) and a 10x higher probe current in the Cs-corrected 603. The detection limit (MMF) using the Cs-corrected probe can be ∼3x better than for the uncorrected HB 603 for the same acquisition conditions. (Reprinted from M. Watanabe, *et al.*, Improvements in the X-Ray Analytical Capabilities of a Scanning Transmission Electron Microscope by Spherical-Aberration Correction, *Microsc. Microanal.* Vol. 12, p. 515–526, Copyright 2006, with permission of Cambridge University Press.)

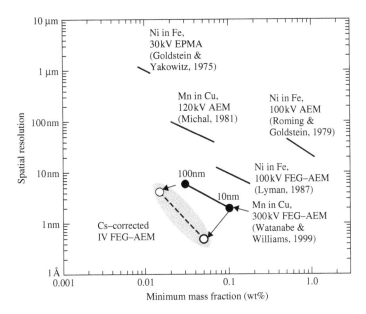

Figure 7.8 A summary of the X-ray analysis performance of several electron-probe instruments with respect to the spatial resolution and the MMF. (Reprinted from M. Watanabe, *et al.*, Improvements in the X-Ray Analytical Capabilities of a Scanning Transmission Electron Microscope by Spherical-Aberration Correction, *Microsc. Microanal.* Vol. 12, p. 515–526, Copyright 2006, with permission of Cambridge University Press.)

during more routine analysis of sample areas. EDX detection in such probe corrected machines are increasingly using arrays of SDDs giving large solid collection angles and efficient collection of the detected signal.

The formation of smaller probes using aberration correction necessarily implies increased probe convergence angles. As discussed in Chapter 6, section 6.6, this means it is essential to collect energy loss electrons over large angles otherwise a large fraction of the available signal is lost. In order for these large collection angles to be coupled optimally into an EEL spectrometer, it is often necessary to add an additional coupling module (post specimen optics) prior to the spectrometer consisting of a set of multipole (e.g. quadrupole and octapole) lenses and auxiliary alignment and adjustment coils (see Figure 6.10). The coupling module can provide an adjustable camera length for subsequent cameras, detectors and spectrometers and also allow the spectrometer entrance crossover height to be adjusted, which is important for aberration correction in the spectrometer itself; the latter increases the collection efficiency of the spectrometer at a given energy resolution.

Finally, source monochromation is really only of benefit for EELS spectral energy resolution, particularly the study of the fine structure in the low loss spectrum and that associated with the ionisation edges in the core loss region; there are also benefits for EFTEM imaging as discussed in section 7.8. However, in virtually all cases, monochromation will result in a significant decrease in beam/probe current (for a given probe size) which will deleteriously affect the MMF (for a given acquisition time).

7.3 ELEMENTAL ANALYSIS AND QUANTIFICATION USING EDX

The identification of the elements contributing to an EDX spectrum, such as that shown in Figure 7.3, is relatively routine. The positions and relative heights of the various X-ray peaks associated with a particular element are most commonly stored in a database associated with the EDX system software package. Working from the high energy end of the spectrum, for internal consistency, it is necessary to confirm (both using autodetection software routines and also manually) that if the K lines for a particular element are present, then the corresponding L (and possibly M lines) are also evident (bearing in mind spectral resolution and possible peak overlap), i.e. if a Ni Kα peak is present then both an Ni Kβ and the series of Ni L X-ray lines must be also observed.

Initial processing of the EDX spectrum prior to quantification may involve the subtraction of a hole count spectrum (discussed in section 7.2.1) to remove the effects of any stray scatter during EDX acquisition from the volume of interest. Subsequent processing of the EDX spectrum then involves not only background subtraction of the underlying Bremsstrahlung X-ray contribution, but correction for both *escape peaks* due to X-ray absorption in the surface of the detector and also *sum peaks* due to the near simultaneous arrival (and hence detection) of two X-rays at high count rates. The exact form of the Bremsstrahlung background will be a function of the detector orientation, the nature and thicknesses of the detector window and detector crystal electrode and dead layers discussed in section 7.2.1. All these procedures may be performed using deconvolution techniques involving Fourier transformation and filtering, or alternatively both the background, escape and sum fractions may be modelled mathematically. Such facilities are an integral part of many currently available EDX software packages.

After the spectrum has been pre-processed, the characteristic X-ray peaks are fitted, using least squares methods, either to stored spectra or to computed profiles. Such a fitting procedure can also deal with the problem of overlapping peaks. Alternatively, simple integration of areas under the peaks may be performed. Once the characteristic line intensities have been extracted the next step is to convert these into a chemical composition present in the irradiated sample volume.

For quantification of the amount of a particular element present, it is necessary to know the relevant cross-section for X-ray excitation by the accelerated electrons as well as the absorption characteristics of the window, the electrode and the dead layer of the X-ray detector, so as to correct the measured X-ray intensity. The most usual approach to tackling this problem, often known as the *Cliff-Lorimer* method after the pioneers (Williams and Carter, 2009), is firstly to deal with solely relative amounts of elements present and secondly, to use a proportionality factor known as a *k-factor* which may either be calculated from first principles or, more usually, can be measured experimentally. The latter approach employs a standard compound of known composition which contains the elements of interest. The basic equation for the analysis of inorganic materials is:

$$C_X/C_Y = k_{XY} \, I_X/I_Y \qquad (7.2)$$

where C denotes the concentration in weight% and I denotes the characteristic X-ray peak intensities above the background, for both elements X and Y. k_{XY} is the appropriate proportionality or *k-factor*, also known as a Cliff-Lorimer sensitivity factor, which is independent of specimen composition and thickness but is dependent on accelerating voltage (E_o). Frequently k-factors are quoted for several element pairs which have a common element such as Si or Fe; k_{XY} is then simply given by $k_{XY} = k_{XSi}/k_{YSi}$. However, Si Kα X-rays are absorbed by EDX detectors and are therefore instrument dependent. Therefore it is then better to choose a common element with characteristic X-rays which are less absorbed by both the detector window and the specimen which will give more instrumentally independent k-factors; metallurgists often employ Fe to circumvent this problem. For elements of $Z \geq 14$, the k-factors determined by several workers are in general agreement. Use of the literature values would result in an error of around $\pm 10\%$. For elements of $Z < 14$ it is inadvisable to use published data, since the variation may be as high as $\pm 50\%$ from instrument to instrument. For greatest accuracy, k-factors can be experimentally determined via measurements on standards of known composition under the same microscope conditions. For the materials scientist, the requirement is a thin foil of known composition containing elements X and Y. Other elements may also be present although these may give to rise to problems with overlapping peaks. Obviously the foil must be stable in the electron beam and homogeneous on the scale of the microanalysis. However, a much more rapid procedure is to use an EDX software packages which can commonly calculate a set of k-factors, obtained for a particular detector system (i.e. with a particular window, electrode and dead layer thickness and a certain geometry relative to the sample) at a given electron beam accelerating voltage and relative to certain element (usually Si); this resulting database is known as a *virtual standards pack*. Briefly, this involves the calculation of the cross-sections for X-ray excitation as well as the detector response function, involving absorption in the window, electrode and the dead layer. Calculation of excitation cross-sections require knowledge of the ionisation cross-sections, the fluorescent yield and the partition factor which governs the distribution of intensity within a particular series of X-ray lines. Overall errors are typically $\pm 15\%$, rising to much higher values for light elements.

A similar equation correlating X-ray intensity and element concentration is used by biologists:

$$C_X = k \, I_X / W_B \qquad (7.3)$$

where C is the concentration expressed as a mass fraction or in molar terms as $mMkg^{-1}$ or, alternatively, as mMl^{-1} and W_B is the intensity of the Bremsstrahlung over a specified energy range which is proportional to thickness for a homogeneous specimen.

For biological analysis, again k is either determined experimentally through the use of standards or it is calculated as before. However, here a standard of known mass thickness, containing the element of interest, in concentration C_{std}, is required. If the thickness of the standard is not the same as that of the specimen, then equation 7.3 needs to be applied twice to give:

$$C_X = C_{std}\{(I_X/W_B)_{unknown}/(I_X/W_B)_{std}\} \qquad (7.4)$$

If the specimen is only a few tens of nanometres in thickness, it is possible to neglect two phenomena: firstly, the differing absorption, within the material itself, of the various characteristic X-rays generated in the specimen as they travel to the specimen exit surface en route to the detector; secondly, fluorescence of one characteristic X-ray by another higher energy characteristic X-ray. However, if the specimen is thicker or we are interested in quantifying the concentration of a light element present in a heavy element matrix then, for accuracy, we have to correct for these effects, in particular differential X-ray absorption. The absorption correction employs an iterative procedure which initially assumes a starting composition for the specimen based on the uncorrected X-ray intensities, the absorption of different energy X-rays emitted at all depths in this specimen en route to the detector is then accounted for using appropriate mass absorption coefficients for X-rays in that material (μ/ρ) and a Beer-Lambert-type expression using the detector sample geometry to calculate the X-ray path length in the sample. The absorption correction then gives a new composition which is subsequently used as input to a further absorption correction. This procedure is repeated until the change in composition falls below some preset level of required accuracy. Normally besides the known collection geometry of the specimen and detector, the only further input parameter for this absorption correction is an estimate of the specimen thickness (t, perhaps obtained by EELS – see section 7.4) and specimen density (ρ). Given the importance of absorption in light element analysis, the need to be aware of the specimen geometry is paramount in order to minimise the X-ray path length through the sample en route to the detector.

It is clear that the magnitude of the absorption effect is a function of μ/ρ, ρ, t and the specimen detector geometry. At any point of analysis all these factors are constant except for μ/ρ which depends on the X-rays being analysed. μ/ρ needs to be the average value for the composition along the path length through the sample to the detector. If it is assumed that the specimen is homogeneous over the path length, then μ can be found from the elemental mass absorption coefficients:

$$(\mu/\rho) = C_X(\mu/\rho)_X + C_Y(\mu/\rho)_Y + \ldots\ldots\ldots \quad (7.5)$$

Although it seems that it is necessary to know the composition before we can calculate μ/ρ, it is often sufficiently accurate to obtain an estimate of the composition from the uncorrected X-ray intensities. If the corrected and uncorrected values differ by more than 10% it may be necessary to recalculate μ/ρ and perform the correction again. Unfortunately, there is not very good agreement between workers for values of μ/ρ for soft X-rays in heavy elements – a common situation for materials scientists.

X-rays causing fluorescence can either be other characteristic X-rays or Bremsrahlung X-rays. The latter are a less significant affect as they tend not to vary greatly from sample to sample and hence tend to cancel out. Fluorescence by characteristic X-rays is only really significant when there are two elements of interest with similar X-ray energies and the energy of one X-ray is just above the absorption edge of another X-ray (e.g. elements close to each other in the periodic table). For K-K fluorescence this means that the atomic numbers of the two elements need to be close in value – typically one apart for low atomic number elements and two apart for medium Z elements. Fluorescence is made considerably worse by the presence of a large concentration of the fluorescing element combined with a small concentration of the fluoresced species. The fluorescence correction factor is determined by the fluorescence ratio, w, the height of the X-ray absorption edge above the background and overvoltage (Jones, 1992).

The combined absorption and fluorescence corrections are usually initiated assuming a starting composition derived from the uncorrected X-ray intensities. A new composition is determined and the results iterated until the result converges to a single solution and a sufficient degree of precision has been achieved.

For accurate analysis it is important to account for the effects of all elements detected in the sample (whether or not the concentrations are required) – both in terms of deconvoluting peak overlaps and also correcting for absorption and fluorescence. It is important to stress that the analyst should think about the experimental setup and the specimen before accepting computer generated results at face value.

Besides the Cliff-Lorimer method, recently a new EDX quantitative method called the ζ (zeta)–factor approach has been developed which incorporates both the absorption correction and also the fluorescence correction (Watanabe and Williams, 2006b) and, in addition to compositions, provides a simultaneous determination of the specimen thickness. ζ–factors can be recorded from pure element standard thin films provided the beam current and hence the electron fluence during EDX acquisition is known.

With careful measurement and analysis, as a general rule of thumb, EDX can detect levels of elements down to about 0.1 atom% with an accuracy of roughly $\pm 5\%$. Figures 7.8 and 7.7 respectively show the STEM/EDX performance of some of the best instruments in terms of the MMF and MDM values as well as the implications of probe correction for their ultimate detection limits. However, the absorption of low energy X-rays (both in the specimen and the detector) becomes a severe problem for elements of $Z < 11$ and this can make light element quantification extremely unreliable without an accurate knowledge of your specimen-detector geometry, the sample thickness and possibly also the use of very carefully and individually determined k-factors or, alternatively, the use of the ζ-factor approach.

7.4 LOW LOSS EELS – PLASMONS, IB TRANSITIONS AND BAND GAPS

Generally the low loss region of the EEL spectrum is dominated by the bulk plasmon excitation and this may be thought of as a resonant oscillation of the valence electron gas of the solid (as pictured in the Drude-Lorentz model for metals) stimulated by the passage of the fast incident electron through the sample. Within the free-electron model of solids an expression for the bulk plasmon energy, E_p, is given by:

$$E_p = h/2\pi \, [Ne^2/(m_e\varepsilon_0)]^{1/2} \qquad (7.6)$$

where N is the valence electron density and ε_0 the permittivity of the free space. This free electron formula works surprisingly well for a range of elements and compounds. Free-electron metals, such as aluminium, show very sharp plasmons, while those in insulators and semiconductors are considerably broader since the valence electrons are damped by scattering with the ion-core lattice which may be incorporated into the free electron theory by use of a characteristic relaxation time for the decay of this plasmon oscillation. Thus, in summary, the energy of the plasmon peak is essentially governed by the density of the valence electrons, and the width by the rate of decay of the resonant mode.

As the plasmon energy and hence plasmon peak position is sensitive to the valence electron density, any changes in this quantity, such as those due to alloying in metals or general structural rearrangement in different microstructural phases, can be detected as a shift in the plasmon energy allowing a means of chemical phase identification. Localised changes in properties such as electrical or thermal conductivity and even elastic modulus are a function of the local valence electron densities, hence determination of the plasmon energy can provide a powerful tool for highly localised property determination in solid microstructures – see for example, Daniels *et al.* (2003a and 2007) who studied a set of pitch-derived graphitising carbons heat treated to different temperatures and hence densities summarised in Figure 7.9.

In a thicker specimen ($\geq 100\,$nm, which is roughly the mean free path for inelastic scattering at $100\,$kV), in addition to the plasmon peak at

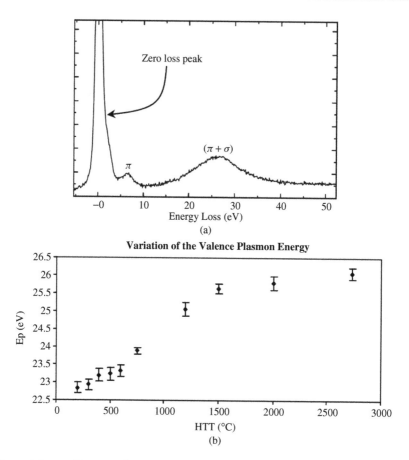

Figure 7.9 (a) The EELS low loss region from a sample of graphitising carbon. The feature at 6.5 eV arises from an interband transition between the π bonding and antibonding orbitals or alternatively a collective plasmon oscillation of π electrons. The bulk valence plasmon (26 eV) arises from a collective oscillation of all the π and σ valence electrons. (b) The variation of bulk valence plasmon energy with sample HTT (heat treatment temperature) (c) The correlation between the valence plasmon energy and the square root of the true (helium) sample density. (Reprinted from H. Daniels, *et al.*, Investigating carbonization and graphitization using electron energy loss spectroscopy (EELS) in the transmission electron microscope (TEM), *Philosophical Magazine*, Vol. 87, p. 4073–4092, Copyright 2008 with permission of Taylor & Francis.)

the energy, E_p, there are additional peaks or harmonics at multiples of the plasmon energy, corresponding to the excitation of more than one plasmon; the intensities of these multiple plasmon peaks follow a Poisson statistical distribution. In the most simplistic form of analysis, the low loss region is just used to determine the relative (or absolute) specimen

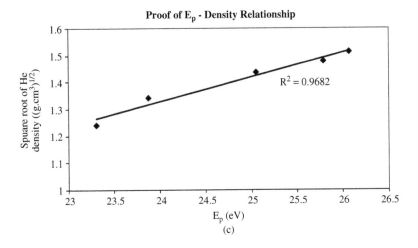

Figure 7.9   (*continued*)

thickness (via the logarithm of the ratio of the total spectral intensity to the zero loss intensity) and to correct EEL spectra for the effects of multiple inelastic scattering when performing quantitative microanalysis on thicker specimens using Fourier Transform-based deconvolution techniques (Brydson, 2001; Egerton, 1996).

In addition to three dimensional bulk plasmons it is also possible to excite two dimensional surface or interface plasmons, these may be apparent in very thin nanolayers as well as small nanoparticles or nanorods; note that these confined plasmon modes are highly dependent on the exact system geometry (e.g. the curvature, radius or aspect ratio of the particle). This general area is known as 'plasmonics' and has benefitted from the reducing energy spread of electron sources including monochromation which allows low energy plasmon signals to be extracted from the zero loss peak.

As well as plasmon oscillations, the low loss region may also exhibit interband transitions, i.e. single electron transitions from the valence band to unoccupied states in the conduction band, which appear as peaks superimposed on the main Lorentzian shape of the plasmon peak. In a solid state physics picture, these interband transitions represent a Joint Density of States (JDOS) – a convolution between the valence and conduction band DOS. The presence of these single-electron excitations can lead to a shift in the energy of the plasmon resonance. In an insulator with a band gap there should no interband transitions below the band gap energy and thus the band gap is associated with an initial rise in

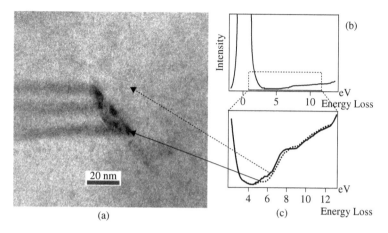

Figure 7.10 (a) STEM BF image of a stacking fault arrangement in CVD single crystalline diamond comprising of primary and secondary faults. (b) EEL spectra taken at locations of the arrows, i.e., in the perfect crystal (grey) and near the stacking fault (black). The difference can be seen more clearly in the expanded view in (c): enhancement in the joint density of states in the vicinity of the partial dislocations bounding the stacking fault occurs due to energy states below the conduction band edge ($\sim$5.5 eV) and due to contribution of sp$_2$ bonding ($\sim$7 eV). (Reprinted from U. Bangert *et al.*, Extended defect related energy loss in CVD diamond revealed by spectrum imaging in a dedicated STEM, *Ultramicroscopy*, Vol. 104, p. 46–56, Copyright 2005 with permission of Elsevier.)

intensity in the low loss region, as seen in Figure 7.10 for the case of diamond (and also a defect in diamond). Accurate removal of the zero loss peak from the rest of the spectrum (which in many cases is not trivial particularly the larger the energy spread of the source) allows the determination of band gap energies and, from the spectral shape of the onset of transitions, whether the gap is direct or indirect, however a more accurate procedure is based on extraction of the dielectric function of the material, described below.

A more sophisticated analysis of the low loss region is based upon the concept of the dielectric function, ε, of the material. This is a complex quantity, dependent on both energy loss and momentum, which represents the response of the entire solid to the disturbance created by the incident electron. The same response function, ε, describes the interaction of photons with a solid and this means that energy loss data, may be correlated with the results of optical measurements in the visible and *uv* regions of the electromagnetic spectrum, including quantities such as refractive index, absorption and reflection coefficients. A major benefit of EELS as a nanocharacterisation tool is that, owing to resolution limitations, optical measurements are by necessity averages of an ensemble of individual nano-objects, whereas low loss EELS data can in principle be obtained from single nanostructures (see Figure 7.11) albeit at a spatial resolution approaching at best 1–2 nm owing to the inherent delocalisation of these low energy excitations (see section 7.8).

7.5 CORE LOSS EELS

7.5.1 Elemental Quantification

Analogous to EDX, since the energy of the EEL ionization edge threshold is determined by the binding energy of the particular electron subshell within an atom – a characteristic value, the atomic type may be easily identified with reference to a tabulated database. The signal under the ionization edge extends beyond the threshold, since the amount of kinetic energy given to the excited electron is not fixed and the intensity or area under the edge is proportional to the number of atoms present, scaled by the cross-section for the particular ionization process, and hence this allows the technique to be used for quantitative analysis. This is achieved by firstly fitting a background (often modelled as a power-law $A.E^{-r}$ where E is the energy loss and A and r are constants) to the spectrum immediately before the edge. This is then subtracted (see Figure 7.12) and the intensity is measured in an energy window, Δ, which begins at the edge threshold and usually extends some 50 to 100 eV above the edge. The next step is to compute the inelastic partial **cross-section**, σ, for the particular inner-shell scattering event under the appropriate experimental conditions, i.e. $\sigma(\alpha, \beta, \Delta, E_0)$; note the bracketed terms in italics simply represent the variables on which the partial cross-section depends. This partial cross-section is calculated for the case of a free atom using simple Hydrogenic or Hartree-Fock-Slater wavefunctions and is generally an integral part of the analysis software package (see Figure 7.12). The measured edge intensity is normalized (i.e. divided) by the partial cross-section so that either different edge intensities can be compared in order to derive a *ratio* of atomic concentrations, or the intensity can be directly interpreted in terms of an *absolute* atomic concentration within the specimen volume irradiated by the electron probe. For the latter case, the measured edge intensity is divided by both the partial cross-section and the combined zero loss and low loss intensity measured over the same energy window, Δ, and under the exact same experimental conditions (e.g. values of convergence and collection angles); the result is usually expressed in terms of an areal density

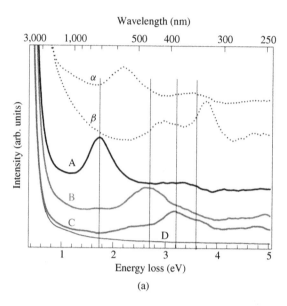

(a)

Figure 7.11 (a) Experimental and simulated EEL spectra. Deconvoluted EEL spectra (solid lines) measured at (A) the corner (B) the edge and (C) the centre of the particle together with (D) the spectrum of the mica support. The last spectrum demonstrates the interest of the choice of a mica support that does not contribute at all to the EEL spectra over the energy domain of interest (1–5 eV). Spectra showing (α) the corner mode of a smaller equilateral triangular nanoprism and (β) the silver bulk mode (3.80 eV) and dipolar surface plasmon mode (3.00 eV) of a quasi-spherical Ag nanoparticle are also represented by dashed lines. The energy of the mode identified at the corners is size-dependent (compare spectra A and α). (b) HAADF–STEM image of the particle, showing the regular geometry that is characteristic of most triangular particles synthesised in this sample. The projected mass image of the scanned region shows the flat top and bottom morphology of the particle (inset). The image contrast around the particle is due to radiation damage in the mica caused by the electron beam. (c) Experimental and simulated EELS amplitude maps EELS amplitude distributions obtained after gaussian fitting of the three modes mapped in. The colour scale, common to the three maps, is linear and in arbitrary units. These maps were obtained by processing a 32×32 spectrum image. In each pixel, the amplitude of a given mode is deduced from the gaussian parameters. This method produces more well-localized spatial distributions as no artificial extension and background blurring of the modes is introduced during the analysis. (Reprinted from J. Nelayah, *et al.*, Mapping surface plasmons on a single metallic nanoparticle, *Nature Physics*, Vol. 3, p. 348–353, Copyright 2007, with permission of Nature Publishing Group.) (A colour version of this figure appears in the plate section)

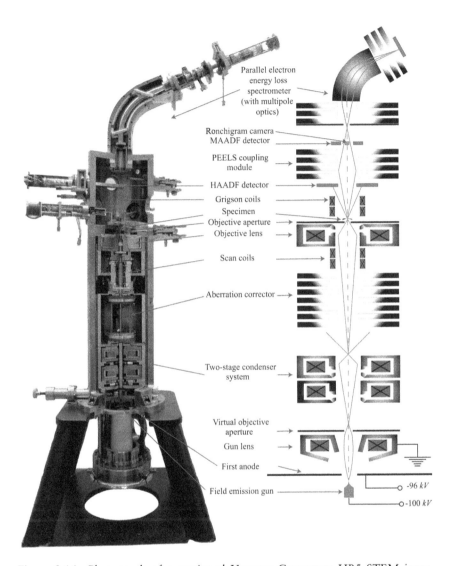

Figure 2.14 Photograph of a sectioned Vacuum Generators HB5 STEM incorporating an early prototype Nion MarkI quadrupole-octupole C_s corrector in the Cavendish Museum plus corresponding schematic diagram. The original microscope, the second one manufactured by VG, was sectioned in the Cavendish Workshop by David Clarke (courtesy: David Clarke, J.J. Rickard and Quentin Ramasse).

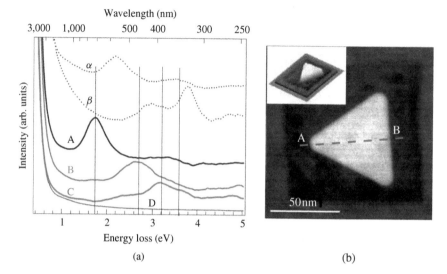

Wavelength (nm)

(a)

(b)

Figure 7.11 (a) Experimental and simulated EEL spectra. Deconvoluted EEL spectra (solid lines) measured at (A) the corner (B) the edge and (C) the centre of the particle together with (D) the spectrum of the mica support. The last spectrum demonstrates the interest of the choice of a mica support that does not contribute at all to the EEL spectra over the energy domain of interest (1–5 eV). Spectra showing (α) the corner mode of a smaller equilateral triangular nanoprism and (β) the silver bulk mode (3.80 eV) and dipolar surface plasmon mode (3.00 eV) of a quasi-spherical Ag nanoparticle are also represented by dashed lines. The energy of the mode identified at the corners is size-dependent (compare spectra A and α). (b) HAADF–STEM image of the particle, showing the regular geometry that is characteristic of most triangular particles synthesised in this sample. The projected mass image of the scanned region shows the flat top and bottom morphology of the particle (inset). The image contrast around the particle is due to radiation damage in the mica caused by the electron beam. (c) Experimental and simulated EELS amplitude maps EELS amplitude distributions obtained after gaussian fitting of the three modes mapped in. The colour scale, common to the three maps, is linear and in arbitrary units. These maps were obtained by processing a 32×32 spectrum image. In each pixel, the amplitude of a given mode is deduced from the gaussian parameters. This method produces more well-localized spatial distributions as no artificial extension and background blurring of the modes is introduced during the analysis. (Reprinted from J. Nelayah, *et al.*, Mapping surface plasmons on a single metallic nanoparticle, *Nature Physics*, Vol. 3, p. 348–353, Copyright 2007, with permission of Nature Publishing Group).

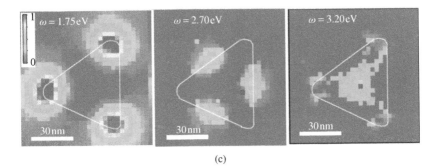

(c)

Figure 7.11 (*continued*)

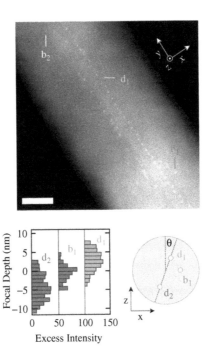

Figure 8.1 Three-dimensional localization of Au atoms. (a) Sum of seven aligned HAADF-STEM images of an intrinsic Si nanowire showing impurities trapped at a twin defect (d1,d2) and bulk impurities (b1,b2). Scale bar is 5 nm. (b) Excess intensity of Au atoms indicated in (a) plotted as a function of focal depth. The peak associated with atom b1 lies between the defect atom peaks and is therefore located within the nanowire as indicated in the schematic diagram. The dashed line extending across the nanowire represents the (111) twin defect bisecting the nanowire. (Reprinted from J. E. Allen, E. R. Hemesath, D. E. Perea, J. L. Lensch-Falk, Z.Y. Li *et al.*, High-resolution detection of Au catalyst atoms in Si nanowires, *Nature Nanotechnology*, Vol. 3, Copyright 2008, with permission of Nature Publishing Group.)

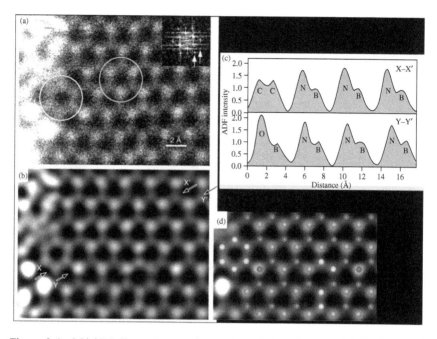

Figure 8.4 MAADF STEM image of monolayer BN. a, As recorded. b, Corrected for distortion, smoothed, and deconvolved to remove probe tail contributions to nearest neighbours. c, Line profiles showing the image intensity (normalised to equal one for a single boron atom) as a function of position in image b along X–X′ and Y–Y′. The elements giving rise to the peaks seen in the profiles are identified by their chemical symbols. Inset at top right in a shows the Fourier transform of an image area away from the thicker regions. Its two arrows point to (1120) and (2020) reflections of the hexagonal BN that correspond to recorded spacings of 1.26 and 1.09Å. Part (d) shows part of a DFT simulation of a single BN layer containing the experimentally observed substitutional impurities overlaid on the corresponding part of the experimental image. Red, B; yellow, C; green, N; blue, O. (Adapted from Krivanek *et al.*, Atom-by-Atom Structural and Chemical Analysis by Annular Dark-Field Electron Microscopy, *Nature*, Vol. 464, p571, Copyright 2010, with permission of Nature Publishing Group.)

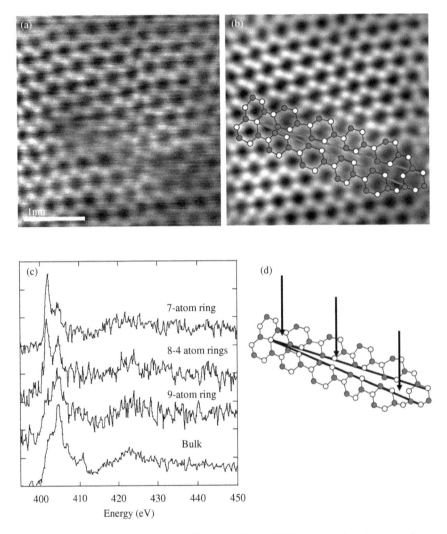

Figure 8.6 *Z*-contrast image of a dissociated mixed dislocation, showing a stacking fault in between the edge and screw dislocation cores (a). The ball and stick model (b) superimposed on a noise filtered copy of (a) clarifies the atomic structure and shows how the stacking fault widens in the plane of the image between the two dislocations to accommodate the strains associated with each dislocation. Other similar partial dislocations connected by stacking faults have been observed with the lengths of the stacking faults extending over 15–30Å in length. Electron-energy loss spectra (c) from a number of positions marked in (d) along the partial dislocation, compared to bulk GaN. The fine structure of the *N K*-edge confirms that one dislocation has a screw character while the other has an edge character. (Adapted from I. Arslan, A. Bleloch, E.A. Stach, and N.D Browning, Atomic and Electronic Structure of Mixed and Partial Dislocations in GaN, *Phys. Rev. Lett.* Vol. 94, p. 25–504, Copyright 2005 with permission of APS.)

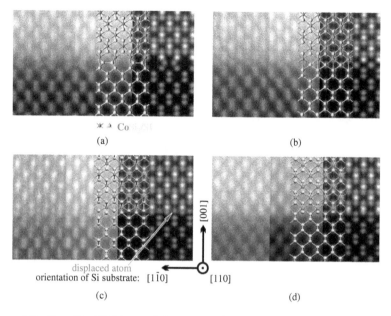

(a)   (b)

orientation of Si substrate: [1$\bar{1}$0]   [110]

(c)   (d)

Figure 8.7 Raw HAADF image (left), averaged contrast (middle), image simulation (right), and ball and stick model (overlaid) for: the (2 × 1) reconstructed, sevenfold interface, showing two projections rotated to each other by 90° around the interface normal, (a) and (b), and the eightfold interface, (c) and (d), showing two projections again rotated. Images were taken in [110] Si substrate zone axis. Averaging over several unit cells along the interface was performed by Fourier filtering with a periodic mask. (Reprinted from M. Falke, *et al.*, Real structure of the $CoSi_2/Si(001)$ interface studied by dedicated aberration-corrected scanning transmission electron microscopy, *App. Phys. Lett.* Vol. 86, p. 203103, Copyright 2005 with permission of American Institute of Physics.)

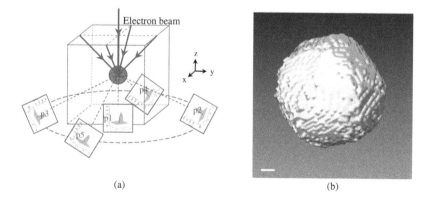

(a)   (b)

Figure 8.13 (a) Schematic showing the five projections needed for the reconstruction and (b) one view of the reconstruction of the 6 nm Pt particle (scale bar 1 nm).

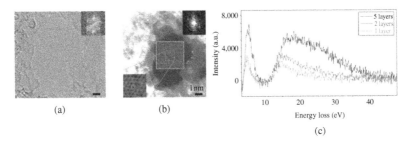

(a)	(b)	(c)

Figure 8.15 High-resolution images of mono-layer graphene. a, b, Bright-field (a) and HAADF (b) images of the monolayer, showing a clean patch of graphene surrounded by a mono-atomic surface layer; individual contaminant atoms of higher atomic number can be seen in b. The inset FFT clearly shows the lattice in the HAADF image and, by applying a bandpass filter, the atomic structure is apparent. c, an EELS spectrum from the monolayer is compared with that from a bi-layer and bulk graphene (graphite); Scale bars in a and b, 1 nm. (Adapted from Gass *et al.*, Free-Standing Graphene At Atomic Resolution, *Nature Nanotechnology*, Vol. 3, p676–681, Copyright 2008, with permission from Nature Publishing Group.)

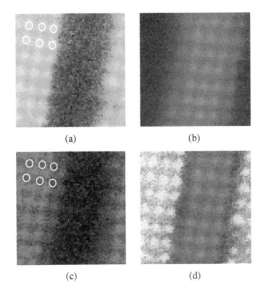

(a)	(b)
(c)	(d)

Figure 8.19 Spectroscopic imaging of a $La_{0.7}Sr_{0.3}MnO_3/SrTiO_3$ multilayer, showing the different chemical sublattices. (a) La M edge; (b) Ti L edge; (c) Mn L edge; (d) red-green-blue false colour image obtained by combining the rescaled Mn, La, and Ti images. Each of the primary colour maps is rescaled to include all data points within two standard deviations of the image mean. The white circles indicate the position of the La columns, showing that the Mn lattice is offset. Live acquisition time for the 64×64 spectrum image was ~30 s; field of view, 3.1 nm. (Reprinted from Muller, D.A., Kourkourtis, L.F., Murfitt, M., Song, J.H., Hwang, H.Y., Silcox, J., Delby, N. and Krivanek, O.L. (2008) Atomic-Scale Chemical Imaging of Composition and Bonding by Aberration-Corrected Microscopy, Science, 319, p1073–1076 with permission of the American Association for the Advancement of Science.)

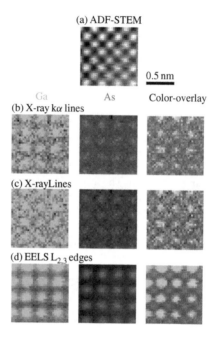

(a) ADF-STEM

0.5 nm

Ga   As   Color-overlay

(b) X-ray kα lines

(c) X-rayLines

(d) EELS L$_{2,3}$ edges

Figure 8.20 (a) HAADF-STEM image of the [001]-projected GaAs simultaneously recorded during SI acquisition; a set of a Ga maps, and As maps and a color-overlay map of K X-ray lines (b), L X-ray lines (c) and EELS $L_{2,3}$ edges (d). (Reprinted from Watanabe, M., Kanno, M. and Okunishi, E. (2010a) Atomic-Resolution Elemental Mapping by EELS and XEDS in Aberration Corrected STEM, *Jeol. News*, 45(1), p 8–15, with permission of Watanabe, M.)

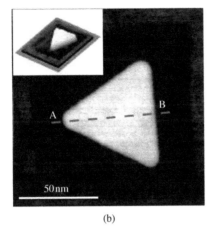

(b)

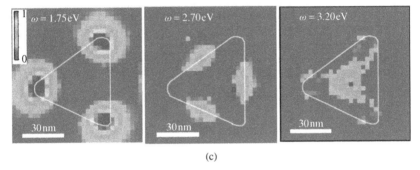

(c)

Figure 7.11   (*continued*)

(in atoms per nm^2) multiplied by the specimen thickness (Brydson, 2001). An example of this may be found in Pan *et al.* (2008) who used STEM/EELS to determine the absolute iron content in ferritin mineral cores within a liver biopsy; a specimen which is discussed further in Chapter 8, section 8.3.4.

Severe problems can arise in EELS detection and elemental quantification if the sample thickness is greater than the mean free path for inelastic scattering, as in this case multiple inelastic scattering will occur. As discussed in the previous section, this will significantly increase the plasmon intensities, leading to an increase in the background contribution making it difficult to identify the presence of edges in a spectrum, a further effect is the transfer of intensity away from the edge threshold towards higher energy losses due to the increase of double scattering events involving a plasmon excitation followed by an ionisation event or vice versa. It is possible to remove this multiple inelastic scattering

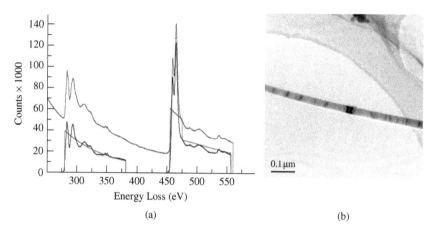

(a)   (b)

Figure 7.12 (a) EELS from a single crystal titanium carbide nanorod (TEM image shown in (b)) shows the EELS carbon K-edge at ca. 282 eV and the titanium $L_{2,3}$-edge at ca. 455 eV both before and after subtraction of the extrapolated background intensities, an integration window of 100 eV is shown. The background subtracted EELS edges are superimposed on the theoretical free atom partial cross-sections. Quantitative analysis gives a Ti/C ratio of 1:1. (Reprinted from X. Li *et al.*, A convenient, general synthesis of carbide nanofibres via templated reactions on carbon nanotubes in molten salt media, *Carbon* Vol. 47, p. 201–208, Copyright 2009, with permission of Elsevier.)

contribution (at the expense of some added noise) from either the whole EELS spectrum or a particular spectral region by Fourier transform deconvolution techniques. Two techniques are routinely employed, one, known as the Fourier-Log method, requires the whole spectrum over the whole dynamic range as input data. This large signal dynamic range can be a problem with data recorded in parallel. The second, known as the Fourier-Ratio method, requires a spectrum containing the feature of interest (i.e an ionisation edge) which has had the preceding spectral background removed. A second spectrum containing the low loss region from the same specimen area is then used to deconvolute the ionisation edge spectrum.

If the sample is sufficiently thin, EELS is particularly sensitive to the detection and quantification of light elements ($Z < 11$) as well as transition metals and rare earths; the analytical sensitivity for these elements being significantly greater than when using EDX and their detection and quantification is generally more reliable. For the remainder of the elements in the periodic table, detection limits for EELS are generally worse than those for EDX and typically lie between 0.1 and 1 atom%. However in favourable circumstances single atom detection has been demonstrated (Figure 8.16 and section 8.3.6). This is principally due to the limited spectral range of the technique relative to EDX as well as the steep and intense background signal upon which the ionisation signal lies. As for EDX, probe correction can improve the values of MMF and MDM achievable by high spatial resolution measurements. Meanwhile, analytical accuracies in elemental quantification using EELS usually lie in the range 5–10% and are principally dependent on the accuracy of the inelastic cross-sections employed as well as the accuracy of the edge background subtraction for a given SNR.

7.5.2 Near-Edge Fine Structure For Chemical and Bonding Analysis

If electrons are scattered via inelastic collisions with the K-shell electrons of free atoms (e.g. such as in the case of a gas) the core loss edges are sharp, saw-tooth like steps displaying no features. Other core-loss excitations in free atoms display a variety of basic edge shapes that are essentially determined by the degree of overlap between the initial- and final-state electronic wavefunctions which may be determined by quantum mechanical calculations of free atoms. In solids, however, the unoccupied electronic states near the Fermi level (the highest occupied electronic energy level) may be appreciably modified by chemical bonding leading to a complex density of states (DOS) and this is reflected in the electron energy loss near-edge structure (ELNES) which lies superimposed on the basic atomic shape within the first 30–40 eV above the edge threshold (see Figure 7.12 which compares the free atom partial cross section with the actual edge shape). The ELNES effectively represents the available electronic states above the Fermi level (see Figure 7.13a), specific to the environment of the atom(s) being ionised, and hence gives information on the local crystallographic structure and chemical bonding (Keast et al., 2001a). Although the exact interpretation is somewhat complex, ELNES can be modelled using electronic structure calculations of the unoccupied DOS using, for instance, density functional theory (DFT) calculations. However, experimentally this DOS is usually a local DOS representative of the environment(s) of the particular atom undergoing ionisation and also usually a DOS of a particular angular momentum symmetry (i.e. s, p, d, f, etc.) relative to the angular momentum of the ionised inner shell owing to the fact that, as is the case for X-ray emission, the optical dipole selection rule operates for the electronic transitions observed in EEL spectra collected with a small collection aperture and hence scattering angle (corresponding to small momentum transfers upon collision). This limits the observed electron transitions at the core loss EEL edges to those in which the angular momentum quantum number, l, changes by ± 1. This results in different edges of the same element probing different symmetries of the final state, i.e. a K-edge will probe the unoccupied p-like DOS, whereas an $L_{2,3}$-edge will probe the unoccupied s- and d-like DOS.

Besides providing a direct means of probing electronic structure, ELNES also allow the determination of simpler chemical concepts such as local coordinations and valence states of atomic species via the use of

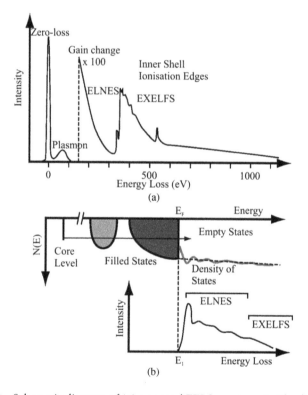

Figure 7.13 Schematic diagram of (a) a general EELS spectrum (with a linear intensity scale and a gain change at high energy loss) showing all of the general observable features. (b) Schematic diagram of the origin of the ELNES intensity which reflects the unoccupied DOS above the Fermi level. (c) Boron K-ELNES fingerprints, measured from the minerals vonsenite (planar-trigonal BO_3 coordination) and rhodizite (tetrahedral BO_4 coordination). (Reprinted from H. Sauer *et al.*, Determination of coordinations and coordination-specific site occupancies by electron energy-loss spectroscopy: an investigation of boron-oxygen compounds, *Ultramicroscopy*, Vol. 49, p. 198–209, Copyright 1993, with permission of Elsevier.)

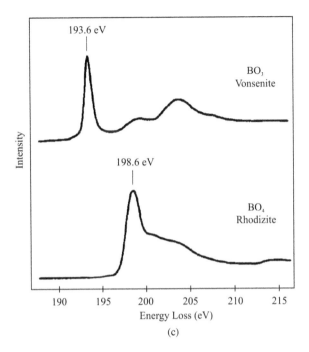

193.6 eV

BO₃
Vonsenite

198.6 eV

BO₄
Rhodizite

Figure 7.13   (*continued*)

'spectral fingerprints' (see Figure 7.13b). In many cases it is found that, for a particular elemental ionisation edge, the observed ELNES exhibits a structure which, principally, is specific to the arrangement, i.e. the number of atoms and their geometry, as well as the type of atoms within solely the first coordination shell. This occurs whenever the local DOS of the solid is dominated by atomic interactions within a molecular unit and is particularly true in many non-metallic systems such as semiconducting or insulating metal oxides where we can often envisage the energy band structure as arising from the broadened molecular orbital levels of a giant molecule. If this is the case, we then have a means of qualitatively determining nearest neighbour coordinations using characteristic ELNES shapes known as coordination fingerprints. A wide range of cations in different coordinations (e.g. aluminium, silicon, magnesium, various transition metals etc.) and anion units (e.g. borate, boride, carbonate, carbide, sulphate, sulphide, nitrate, nitride, etc.) in inorganic solids show this behaviour which can be of great use in phase identification and local structure determination (Garvie *et al.*, 1994; Keast *et al.*, 2001a; Ahn, 2004).

An additional chemical use of ELNES is for the spatially resolved determination of the formal valency of elements in a solid (Calvert et al., 2005). The valence or oxidation state of the particular atom undergoing excitation influences the ELNES in two distinct ways. Firstly, changes in the effective charge on an atom leads to shifts in the binding energies of the various electronic energy levels (both the initial core level and the final state) which often manifests itself in an overall chemical shift of the edge onset. As previously mentioned, absolute energy calibration in EELS has to date been somewhat difficult due to the large spectral dynamic range and the difficulty in recording both the low loss and high loss regions near simultaneously although recent developments in 'dual EELS' should overcome this (Gubbens et al., 2010). Secondly, the valence of the excited atom can affect the intensity distribution in the ELNES. This predominantly occurs in edges which exhibit considerable overlap between the initial and final states and hence a strong interaction between the core hole and the excited electron leading to the presence of quasiatomic transitions (so called since the observed ELNES is essentially atomic in nature and only partially modified by the crystal field due to the nearest neighbour atoms). Examples of such spectra are provided by the $L_{2,3}$-edges of the 3d and 4d transition metals and their compounds and the $M_{4,5}$-edges of the rare earth elements. These spectra exhibit very strong, sharp features known as white lines which result from transitions to energetically narrow d or f bands. This makes detection and quantification of these elements extremely easy. Such spectra may be modeled using atomic multiplet theory in the presence of a crystal field of the appropriate ligand field symmetry (see Figure 7.14 and Keast et al., 2001a).

For certain edges, as well as for certain compounds, the concept of a local coordination or valence fingerprint breaks down. In these cases, the unoccupied DOS accessible to the excited electron cannot be so simply described on such a local level and ELNES features are found to depend critically on the type and arrangement of the atoms in outerlying coordination shells allowing medium range structure determination. An example of this behaviour is provided by the case of the carbon K-edge of cubic titanium carbide in Figure 7.12. This opens up possibilities for the differentiation between different structural polymorphs, as well as the accurate determination of lattice parameters, vacancy concentrations and substitutional site occupancies in complex structures (Scott et al., 2001). In anisotropic crystalline structures orientation-dependent measurements are also possible (Daniels et al., 2003b).

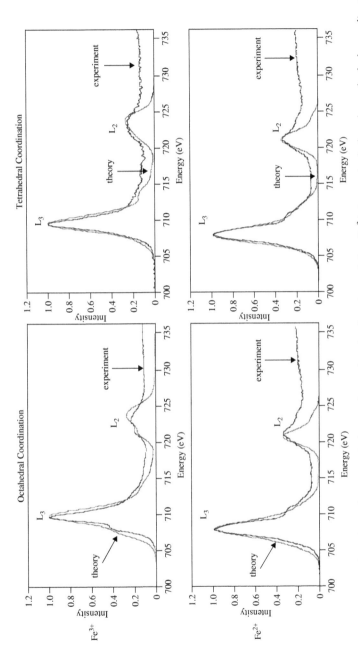

Figure 7.14 Fe $L_{2,3}$-edges from the minerals haematite and orthoclase (containing Fe^{3+} in octahedral and tetrahedral coordination respectively), and hedenbergite and hercyanite (both containing Fe^{2+} in octahedral and tetrahedral coordination respectively); labels indicate experimental EELS results and theoretical modelling using atomic multiplet theory (Keast *et al.*, 2001a). Note the shift to higher energy (from 707.5 eV to 709.5 eV) with increasing oxidation state of Fe. Theoretical data calculated by Derek Revill and Andrew Scott, University of Leeds. (Reprinted from R. Brydson, *Electron energy-loss spectroscopy and energy dispersive X-ray analysis*, *Nanocharacterisation* (Eds. A.I. Kirkland and J.L. Hutchinson), p. 94–137, Copyright 2007, with permission of the Royal Society of Chemistry.)

7.5.3 Extended-Edge Fine Structure For Bonding Analysis

Beyond the near-edge region, superimposed on the gradually decreasing tail of the core loss edge, a region of weaker, extended oscillations are observed, known as the extended energy loss fine structure (EXELFS). In terms of scattering theory, the main distinction between the ELNES and EXELFS regions lies in the fact that the low kinetic energy of the ejected electron in the near-edge region means that it samples a greater volume (the mean free path is large) and multiple elastic scattering occurs – so providing geometrical information on bonding arrangements. In the EXELFS regime, the higher kinetic energy results in predominantly single elastic scattering of the ejected electron – so giving short-range information.

Analysis of this region may be used to extract local structural information such as the length of chemical bonds. There are two approaches to bond length determination. The first method is to analyse the weak EXELFS oscillations occurring some 40–50 eV above the edge onset. As in the more commonly known X-ray synchrotron technique of EXAFS, the period of the oscillations may be used to determine bond distances, while the amplitude reflects the co-ordination number of the particular atom. Since these oscillations are weak, high statistical accuracy (i.e. high count rates and long acquisition times) is required if useful information is to be extracted (Ahn, 2004). The second procedure employs the energy position of the broad ELNES peaks some 20–30 eV above the edge onset, known as multiple scattering resonances (MSR). As their name suggests, these features arise from a resonant scattering event involving the excited electron and a particular shell of atoms. The energies of these features above the edge onset have been shown to be proportional to $1/R^2$, where R is the bond length from the ionised atom. Identification of such MSR permits a semi-quantitative determination of nearest neighbour, and in some cases second nearest neighbour bond lengths (Kurata *et al.*, 1993; Daniels *et al.*, 2007).

At larger scattering angles the dipole selection rules break down and other transitions are observed. At very large angles, a new regime is encountered in which the electrons in the sample may be regarded as if they were free and a hard sphere collision occurs with an associated large momentum transfer resulting in the Electron Compton profile. The width of this feature represents a Doppler broadening of the energy of the scattered electrons due to the initial state momentum of the electrons in the sample and can, in principle, also give bonding information (Egerton, 1996; Brydson, 2001). However, little work has been done in this area for a number of years.

7.6 EDX AND EELS SPECTRAL MODELLING

7.6.1 Total Spectrum Modelling

One approach to the quantitative analysis of spectra is via the fitting of a model to the experimental observations with respect to a certain set of parameters (which are both microscope/spectrometer, as well as sample dependent) using a criterion of goodness of fit, such as, least squares, least absolute values or maximum likelihood. The outcome is a set of parameters which give rise to a 'best fit'.

This approach has been used for EDX spectra for some years principally through the *Desk Top Spectrum Analyser (DTSA)* software package originating from the National Institute for Standards and Technology (NIST) which allows EDX spectra to be manually simulated for a given set of experimental parameters (DTSA 2010). In recent years this has also been attempted for the more complicated case of EEL spectra, see for instance *EELS Model*, a software package to quantify EEL spectra by using model fitting (EELS model 2010 by J. Verbeeck)

7.6.2 EELS Modelling of Near Edge Structures and also the Low Loss

As has been mentioned in section 7.5.2, using density functional band structure calculations (or their equivalents) it is possible to model the occupied and unoccupied DOS and compare this directly with the fine structure displayed in an EEL spectrum. The electron energy bands need to be integrated into a density of electronic states which need to be resolved into a site projection in the unit cell for the atom undergoing ionisation and also an angular momentum symmetry projection for the exact final states which are accessed by the excited electron under the dipole selection rule. Currently a number of different approaches are commonly used, amongst which are:

- WIEN2K (http://www.wien2k.at; Schwarz *et al.*, 2002);
- CaSTEP (http://www.castep.org/; Clark *et al.*, 2005);
- FEFF (http://leonardo.phys.washington.edu/feff/; Ankudinov *et al.*, 1998),

all of which now have specific modules to enable output of EELS spectra. Wien2k is what is known as an all electron code and expands the electron wavefunctions as plane waves, whereas CaSTEP is a pseudopotential code, approximating tightly bound inner core levels as part of the potential term so allowing more complex systems to be handled. Both codes perform calculations in reciprocal space and thus require a periodic system (i.e. a unit cell). Meanwhile FEFF calculates electronic states by considering multiple (elastic) scattering of the excited electron with a real space cluster which more easily allows the study of non-periodic structures.

For comparison with ELNES, since the initial state of the electron transition is an inner core level, this has a very narrow energy width and one only needs to calculate the relevant unoccupied projected DOS above the Fermi level. This has been the most common approach (Keast *et al.*, 2001a; Scott *et al.*, 2001) and an example is shown in Figure 7.15. One major issue, in calculations of some materials, is the inclusion of the core hole produced during the EEL excitation process and its effect on the resultant DOS (Seabourne *et al.*, 2010).

For the case of the low loss, the initial state is no longer an atomic-like, inner shell and one therefore need to perform a more complicated convolution between the occupied valence band DOS and the unoccupied conduction band DOS (a joint density of states) to reproduce the low energy loss spectrum. This has been achieved in recent years (Keast, 2005; Barnes *et al.*, 2007). One issue here is that DFT calculations, particularly when using the so-called local density approximation (LDA) for the difficult parts of the potential (the exchange and correlation of the electrons), rarely produce an accurate measure of the band gap between the valence and conduction bands. Future developments may involve the use of time dependent DFT.

In terms of an aberration corrected probe with a high convergence angle (e.g. 50 mrads), in section 7.2.3.2 it was mentioned that in order not to lose a significant fraction of the energy loss signal, extremely large collection angles are required particularly for high energy losses. This may mean that we are in a range of momentum transfer where non-dipole transitions could be evident (e.g. quadrupole transitions where the change in angular momentum quantum number is two, such as s to d-like transitions observed at $L_{2,3}$-edges) which may result in subtle changes in ELNES.

7.7 SPECTRUM IMAGING: EDX AND EELS

Besides simple point analysis, in STEM, it is possible to raster the electron probe across the specimen and record an EDX or EELS spectrum at every specimen pixel (x,y) – this technique being generally termed

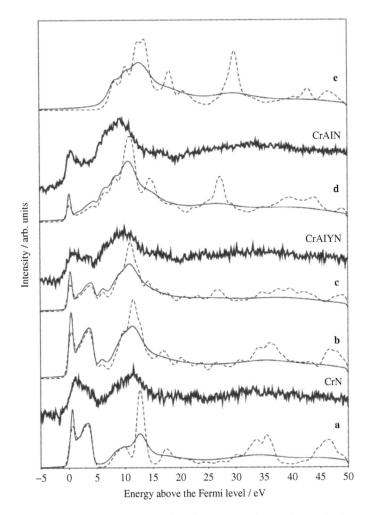

Figure 7.15 Comparison of the simulated ELNES and experimental observations for: (a) NCr (b) NCr$_{0.75}$Al$_{0.25}$ (c) NCr$_{0.5}$Al$_{0.5}$ (d) NCr$_{0.25}$Al$_{0.75}$ (e) cubic NAl. In all cases the solid black line includes energy-dependent final-state lifetime broadening, the dashed line is with minimal numerical broadening. The experimental results for CrN, CrAlYN and CrAlN are shown with solid bold lines as labelled. (Reprinted from I.M. Ross, *et al.*, Electron energy loss spectroscopy of nano-scale CrAlYN/CrN-CrAlY(O)N/Cr(O)N multilayer coatings deposited by unbalanced magnetron sputtering, *Thin Solid Films*, Vol. 518, p. 5121-5127, Copyright 2010, with permission of Elsevier.)

'Spectrum Imaging'. The complete dataset may then be processed (either off- or increasingly on-line) to form a quantitative one dimensional linescan or two dimensional map of the sample using either standard elemental quantification procedures for elemental composition using either EDX (see Figure 7.16, Keast *et al.*, 2001b) and/or EELS outlined previously, or the position and/or intensity of characteristic low loss or ELNES features so as to obtain linescans or maps related to variations in chemical bonding. The inherent high spatial resolution of the EELS spectrum imaging technique has allowed maps of elemental distributions to be formed at sub-nanometre and even atomic resolution. Indeed *the first* demonstration of atomic resolution EELS mapping (see Chapter 8, Figure 8.18) was achieved with data recorded from SuperSTEM (Bosman *et al.*, 2007); this has been subsequently also demonstrated by other groups (e.g. Muller *et al.*, 2008 shown in Figure 8.19). Future

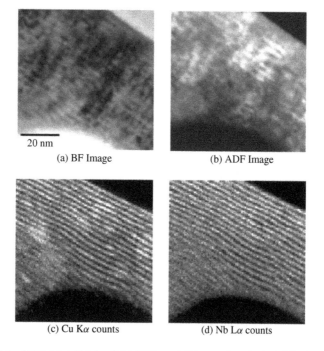

20 nm

(a) BF Image   (b) ADF Image

(c) Cu Kα counts   (d) Nb Lα counts

Figure 7.16   STEM (a) Bright field (BF) and (b) annular dark field (ADF) images from a 1.5 nm Cu-Nb multilayer. EDX Elemental maps acquired using the (c) Cu Ka peak and (d) Nb La peaks. (Reprinted from V.J. Keast, A. Misra, H. Kung and T.E. Mitchell, Compositional distributions in nanoscale metallic multilayers studied using x-ray mapping, *Journal of Materials Research*, Vol. 16, p. 2032–2038, Copyright 2001, with permission of Cambridge University Press.)

developments will include the production of atomic resolution chemical maps using variations in ELNES. Recently the possibility of near atomic resolution EDX maps has also been demonstrated (Watanabe, 2009 shown in Figure 8.20), despite the inherent problems of probe broadening in the specimen.

Recent developments in both EELS and EDX spectrum imaging have included the processing of data using Multi-Variate Statistical Analysis (MSA) methods (such as Principal Component Analysis) followed by the removal of the components identified as noise and then the subsequent regeneration of a noise-reduced dataset (Watanabe, 2009). This processed dataset improves element detectability and quantification as well as identification of changes in ELNES as a function of spatial coordinate. More recent developments in EELS spectrum imaging involve fully computerised beam control and beam blanking which allows complex, discontinuous areas in an image to be scanned during the acquisition of a spectrum image. As discussed in section 6.4.5, this is commonly termed 'smart acquisition' (Sader et al., 2010) and allows pre-selected regions within a microstructure (e.g. particle surfaces) to be interrogated and, if necessary, averaged at extremely low electron fluences, so avoiding radiation damage in beam sensitive structures which is a major and often ignored issue in high spatial resolution analysis as is discussed in Chapter 1, section 1.5.

Finally, STEM/EELS spectrum imaging is directly comparable to direct energy filtered imaging in the CTEM (EFTEM) (e.g. Chapter 6 in Ahn, 2004) which involves the selection of a specific energy loss value, or narrow range of energy losses ('an energy window'), from the transmitted electron beam via use of energy selecting slit after the EEL spectrometer. This energy filter is then combined with subsequent image forming optics. The spectrometer and post-spectrometer image forming system may form part of a post column imaging filter (e.g. as manufactured by the company Gatan) or the microscope system may employ an in-column design with the EELS spectrometer and slit placed between the objective and projector lenses (the so-called Ω filter as manufactured by Carl Zeiss and also JEOL). Generally, EFTEM is better suited to mapping from larger fields of view with relatively short acquisition times than STEM/EELS spectrum imaging. The spatial resolution of EFTEM is limited to about 1 nm or so which is poorer than that of EELS spectrum imaging, particularly since the development of aberration corrected probes. However there are certain limitations to the ultimate spatial resolution of EELS which are discussed in the next section. If these limitations can be addressed by the choice of the optimum experimental conditions, then the use of STEM/EELS spectrum imaging with aberration corrected probes (and even the use of EFTEM with CTEM image correctors – see Chapter 9 and particularly Figure 9.1) can produce atomically resolved elemental or even chemical images.

7.8 ULTIMATE SPATIAL RESOLUTION OF EELS

In addition to broadening of the probe by elastic scattering, possible channelling and the reduction in effective broadening by use of an EELS angle limiting collector aperture (see section 7.2), there is another factor, termed delocalisation, which contributes to the spatial resolution of EELS which is independent of specimen thickness and is inherent to the inelastic scattering process itself. As discussed in section 3.4.4 and elsewhere, the delocalisation distance of an EEL signal can be regarded as a wave-optics resolution limit owing to the limited angular range of inelastic scattering; typically it is a few nm for plasmon scattering dropping to 0.5 nm for losses greater than 100 eV (Egerton, 2007).

Furthermore, for any inelastic scattering process such as EELS (or EDX) it is also highly important to consider the complicating effects of the highly convergent (aberration corrected) STEM probe, the elastic dynamical scattering of the fast electron in the specimen as well as the specific geometry of the detector. For the case of EELS data measured with atomic scale aberration corrected probes, owing to the large probe convergence angle, then if the spectrometer collection angle cuts off an appreciable part of the inelastic scattering (section 6.6), then delocalisation effects can become much more complex (Egerton, 2009). The inelastic cross section can then be rewritten in terms of an effective non-local potential which is related to a quantity called the mixed dynamic form factor which attempts to describe the coupling between elastic and inelastic scattering (Witte et al., 2009). The EELS scattering cross section now includes the effect of both the probe and the detector geometry (and is therefore a function of two spatial coordinates). This formulation essentially incorporates the phase of the transmitted electron and its elastic interaction (i.e. diffraction) with the specimen resulting in an inelastic cross-section that varies sensitively with angle between the electron trajectory and the crystal.

These effects have highlighted two main issues: firstly atomic resolution STEM/EELS spectrum images of elemental distributions may not always be directly visually interpreted, i.e. the inelastic intensity in the

EELS spectrum image may not by directly proportional to the integrated current density and hence apparent atom positions in atomic resolution EELS maps may not necessarily correspond to regions of high intensity in atomically resolved HAADF images. The latter is particularly true of EELS spectrum image maps recorded with aberration corrected STEMs and a relatively small collection angle (relative to the probe convergence angle). As a general rule, delocalisation in EELS spectrum images decreases with: (a) higher energy loss and lower acceleration voltages (which decreases the effective scattering angle associated with the energy loss event); (b) specimen thickness; (c) reduced channelling (i.e. tilt away from major zone axis); and (d) increasing EELS collection angle (in order to collect as much as possible of the inelastic signal produced by the converged probe, then generally for STEM/EELS with a converged probe, the collection angle needs to be at least twice the convergence angle). For EFTEM, as well as the above factors, increased coherency in the electron source helps to localise the signal ideally the use of a monochromated source and an EELS imaging filter corrected for non-isochromaticity (i.e. differences in energy at different positions within an EFTEM image due to aberrations in the filter).

Secondly, in terms of modelling EELS spectra, particularly atom-column resolved ELNES calculations, the situation with atomic scale probes may mean that it is not simply a case of calculating the probe intensity as a function of depth (through thickness z) and combining this intensity distribution with site and symmetry resolved DFT calculations. A revised theory for modelling ELNES is then required which incorporates the effects of this non local potential on the electron probe wavefunction (Witte *et al.*, 2009).

7.9 CONCLUSION

This chapter has summarised the basic aspects of the excitation mechanisms, the interpretation, the acquisition and processing of EDX and EELS spectra. In addition we have attempted to highlight the benefits (and in some cases complications) introduced by the use of small, highly converged, aberration corrected probes. In principle, when combined with aberration correction (particularly probe correction), both techniques offer the potential for truly atomically resolved spectroscopy of nanostructured solids. Hence we expect considerable progress to be made in this field of research in future years.

REFERENCES

Ahn, C.C. (ed.) (2004) *Transmission EELS in Materials Science*, 2nd edn, John Wiley & Sons: Weinheim.

Ahn, C.C. and Krivanek, O.L. (1983) EELS Atlas, Gatan Inc./HREM Facility Arizona State University, Warrendale PA.

Ankudinov, A.L., Ravel, B., Rehr, J.J. and Conradson, S.D. (1998) Real Space Multiple Scattering Calculation of XANES, *Phys. Rev. B.*, 58, 7565–7576. See also http://leonardo.phys.washington.edu/feff/.

Bangert, U., Harvey, A.J., Schreck, M. and Hörmann, F. (2005) Extended Defect Related Energy Loss in CVD Diamond Revealed by Spectrum Imaging in a Dedicated STEM, *Ultramicroscopy* 104, 46–56.

Barnes, R., Bangert, U. and Scott, A. (2007) Investigating Large Valency Clusters in Type IIa Diamond With Electron Energy Loss Spectroscopy, *Phys. State. Sol.* 204, 3065–3071.

Bosman, M., Keast, V.J., García-Muñoz, J.L., D'Alfonso, A.J., Findlay, S.D. and Allen, L.J. (2007) Two-Dimensional Mapping of Chemical Information at Atomic Resolution, *Phys. Rev. Lett.* 99, 086102.

Brydson, R. (2001) *Electron Energy Loss Spectroscopy*, Bios: Oxford.

Calvert, C.C., Brown, A. and Brydson, R. (2005) Determination of the Local Chemistry of Iron in Inorganic and Organic Materials, *J. Electron Spectrosc. Relat. Phenom.*, 143, 173.

Clark, S.J., Segall, M.D., Pickard, C.J., Hasnip, P.J., Probert, M.J., Refson, K. and Payne, M.C. (2005) First Principles Methods Using CASTEP. *Z. Kristallogr.*, 220, 567–570. See also http://www.castep.org/.

Daniels, H., Brown, A., Brydson, R. and Rand, B. (2003a) Quantitative Valence Plasmon Mapping in the TEM: Viewing Physical Properties at the Nanoscale, *Ultramicroscopy*, 96, 547–558.

Daniels, H., Brown, A., Scott, A., Nichells, T., Rand, B. and Brydson, R. (2003b) Experimental and Theoretical Evidence for the Magic Angle in Transmission Electron Energy Loss Spectroscopy, *Ultramicroscopy*, 96, 523–534.

Daniels, H., Brydson, R., Rand, B. and Brown, A. (2007) Investigating Carbonization and Graphitization Using Electron Energy Loss Spectroscopy (EELS) in the Transmission Electron Microscope (TEM), *Philos. Mag.*, 87, 4073–4092.

DTSA (2010) – see http://www.cstl.nist.gov/div837/Division/outputs/DTSA/DTSA.htm.

EELS model (2010) see http://www.eelsmodel.ua.ac.be.

Egerton, R.F. (1996) *Electron Energy Loss Spectroscopy in the Electron Microscope*, Plenum Press: New York.

Egerton, R.F. (2007) Limits to the Spatial, Energy and Momentum Resolution of EELS, *Ultramicroscopy*, 107, 575–586.

Egerton, R.F. (2009) Electron Energy Loss Spectroscopy in the TEM, *Rep. Prog. Phys.*, 72, 016502.

Garvie, L.A.J., Craven, A.J. and Brydson, R. (1994) Use Of Electron-Energy-Loss Near-Edge Fine-Structure In The Study Of Minerals, *Am. Mineral.*, 79, 411–425.

Gubbens, A., Barfels, M., Trevor, C., Twesten, R., Mooney, P., Thomas, P., Menon, N., Kraus, B., Mao, C. and McGinn, B. (2010) The GIF Quantum, A Next Generation Post-Column Imaging Energy Filter, *Ultramicroscopy*, 110, 962–970.

Herzing, A.A., Watanabe, M., Edwards, J.K., Conte, M., Tang, Z.R., Hutchings, G.J. and C.J. Kiely (2008) Energy Dispersive X-ray Spectroscopy of Bimetallic Nanoparticles in an Aberration Corrected Scanning Transmission Electron Microscope, *Faraday Discussions*, 138, 337–351.

Jones, I.P. (1992) *Chemical Microanalysis using Electron Beams*, Institute of Materials: London.

Keast, V.J., Scott, A.J., Brydson, R., Williams, D.B. and Bruley, J. (2001a) Electron Energy Loss Near Edge Structure – a Tool for the Investigation of Electronic Structure on the Nanometre Scale, *J. Microscopy*, 203, 135–175.

Keast, V.J., Misra, A., Kung, H. and Mitchell, T.E. (2001b) Compositional Distributions in Nanoscale Metallic Multilayers Studied Using X-ray Mapping, *Journal Mater. Res.*, 16, 2032–2038.

Keast, V.J. (2005) Ab Initio Calculations of Plasmons and Interband Transitions in the Low-Loss Electron Energy-Loss Spectrum, *J. Electron Spectrosc. Relat. Phenom.*, 143, 97.

Kurata, H., Lefevre, E., Colliex, C. and Brydson, R. (1993) Electron-Energy-Loss Near-Edge Structures In The Oxygen K-Edge Spectra Of Transition-Metal Oxides *Phys. Rev. B.*, 47(20): 13763–13768.

Li, X., Westwood, A., Brown, A., Brydson, R. and Rand, B. (2009) A Convenient, General, Synthesis of Carbide Nanofibres Via Templated Reactions On Carbon Nanotubes in Molten Salt Media, *Carbon*, 47, 201–208.

Muller, D.A., Fitting Kourkoutis, L., Murfitt, M., Song, J.H., Hwang, H.Y., Silcox, J., Dellby, N. and Krivanek, O.L. (2008) Atomic-Scale Chemical Imaging of Composition and Bonding by Aberration-Corrected Microscopy, *Science*, 319, 1073–1076.

Nelayah, J., Kociak, M., Stéphan, O., García de Abajo, F.J., Tencé, M., Henrard, L., Taverna, D., Pastoriza-Santos, I., Liz-Marzán, L.M. and Colliex, C. (2007) Mapping Surface Plasmons On a Single Metallic Nanoparticle, *Nature Physics*, 3, 348–353.

Pan, Y.H., Brown, A., Sader, K., Brydson, R., Gass, M. and Bleloch, A. (2008) Quantification of Absolute Iron Content in Mineral Cores of Cytosolic Ferritin Molecules in Human Liver, *Mater. Sci. Tech. – Lond.* 24(6): 689–694.

Ross, I.M., Rainforth, W.M., Seabourne, C.R., Scott, A.J., Wang, P., Mendis, B.G., Bleloch, A.L., Reinhard, C. and Hovsepian, P.E. (2010) Electron Energy Loss Spectroscopy of Nano-Scale CrAlYN/CrN-CrAlY(O)N/Cr(O)N Multilayer Coatings Deposited by Unbalanced Magnetron Sputtering, *Thin Solid Films*, 518, 5121–5127.

Sader, K., Schaffer, B., Vaughan, G., Brydson, R., Brown, A. and Bleloch, A. (2010) 'Smart Acquisition EELS', *Ultramicroscopy*, 110, 998–1003.

Sauer, H., Brydson, R., Rowley, P.N., Engel, W. and Thomas, J.M. (1993) Determination of Coordinations and Coordination-Specific Site Occupancies By Electron Energy-Loss Spectroscopy: An Investigation of Boron-Oxygen Compounds, *Ultramicroscopy*, 49, 198–209.

Schwarz, K., Blaha, P. and Madsen, G.K.H. (2002) Electronic Structure Calculations of Solids Using the WIEN2k Package for Material Sciences., *Comp. Phys. Comm.*, 147, 71–76. See also http://www.wien2k.at/.

Scott, A.J., Brydson, R., MacKenzie. M. and Craven, A.J. (2001) A Theoretical Investigation of the ELNES of Transition Metal Carbides and Nitrides for the Extraction of Structural and Bonding Information, *Phys. Rev. B.*, 63, 245105.

Seabourne, C.R., Scott, A.J., Vaughan, G., Brydson, R., Wang, S.G., Ward, R.C., Wang, C., Kohn, A., Mendis, B. and Petford-Long, A.K. (2010) Analysis Of Computational EELS Modelling Results For MgO-Based Systems, *Ultramicroscopy*, 110, 1059–1069.

Walton, A.S., Allen, C.S., Critchley, K., Gorzny, M.L., McKendry, J.E., Brydson, R., Hickey, B.J. and Evans, S.D. (2007) Four-Probe Electrical Transport Measurements on Individual Metallic Nanowires, *Nanotechnology*, 18, 065204.

Watanabe, M., Ackland, D.W., Burrows, A., Kiely, C.J., Williams, D.B., Krivanek, O.L., Dellby, N., Murfitt, M.F. and Szilagyi, Z. (2006a) Improvements in the X-ray Analytical Capabilities of a STEM by Spherical Aberration Correction, *Microscopy and Microanalysis*, 12, 515–526.

Watanabe, M. and Williams, D.B. (2006b) The Quantitative Analysis of Thin Specimens: a Review of Progress From the Cliff-Lorimer to the New ζ factors Methods, *Journal of Microscopy*, 221, 89–109.

Watanabe, M. (2009) Atomic-Resolution Chemical Analysis by Electron Energy-Loss Spectrometry and X-ray Energy Dispersive Spectrometry in Aberration-Corrected Electron Microscopy, *Atomic Level Characterization for New Materials and Devices '09 (ALC09), The 141st Committee on Microbeam Analysis*, Japan Society for the Promotion of Science, pp. 400–405.

Williams, D.B. and Carter, C.B. (2009) *Transmission Electron Microscopy: A Textbook for Materials*, 2nd edn., Springer.

Witte, C., Findlay, S.D., Oxley, M.P. and Rehr, J.J. (2009) Theory of Dynamical Scattering in Near-Edge Electron Energy Loss Spectroscopy, *Phys. Rev. B.*, 80, 184108.

8

Applications of Aberration-Corrected Scanning Transmission Electron Microscopy

Mervyn D. Shannon

SuperSTEM Facility, STFC Daresbury Laboratories, Daresbury, Cheshire, UK

8.1 INTRODUCTION

The advent of aberration-corrected STEM has enabled a number of experiments that were previously impossible. These are partly due to the smaller probe giving improved image resolution and localisation of atom positions, but also the order of magnitude higher current arising from an increase in probe convergence angle by a factor of three to five times makes possible spectroscopy experiments at the scale of an atomic column with sufficient signal to noise ratio and hence detectability in terms of the minimum mass fraction and minimum detectable mass as discussed in Chapter 7.

Whilst bright field imaging is still a useful tool, high-angle annular dark-field (HAADF) becomes the imaging mode of choice for reasons discussed in Chapter 3. For lighter elements a medium-angle dark-field

Aberration-Corrected Analytical Transmission Electron Microscopy, First Edition.
Edited by Rik Brydson.
© 2011 John Wiley & Sons, Ltd. Published 2011 by John Wiley & Sons, Ltd.

(MAADF) detector (see section 6.9.4) is used because these scatter to lower angles. On some instruments there can also be a secondary electron and/or a back-scatter detector for surface/near surface imaging.

Because it is less susceptible to beam-spreading and is often more efficient in terms of signal collection, as discussed in Chapter 7, EELS is generally the preferred analysis tool rather than EDX in aberration-corrected STEM instruments. Nonetheless, for EELS analysis from areas of the order of 0.5 nm in a sample just 10 nm thick, the larger convergence angles required to form smaller probes will lead to significant beam spreading from simple geometrical considerations alone (see section 6.2.9). However, there are clearly good reasons for having EDX as well as EELS in an analytical aberration-corrected STEM – for example the greater coverage of elements across the Periodic Table and convenience for element identification, and more recently the increase in solid angle of detection possible with arrays of SDD detectors (Silicon Drift Detectors). Also the very thin samples required to make the most of aberration-corrected STEM, generally a few tens of nanometres thick, limit the beam spreading due to scattering within the sample and enhance the spatial resolution of EDX analysis.

The examples described below are meant simply as illustrations of the breadth of materials' problems that can be addressed and are not intended to be a comprehensive review. One danger of ever greater detail in imaging and analysis is that you lose sight of the bigger picture. For that reason it is our philosophy that problems tackled in aberration-corrected STEM instruments should have first been thoroughly explored in conventional instruments.

8.2 SAMPLE CONDITION

Samples for Aberration-Corrected STEM must crucially be very clean. Thin layers of amorphous material due to polishing or careful FIB sectioning can often be accommodated. Their main influence is to lower the contrast in the image (Mkhoyan et al., 2008), provided the beam is focused at the interface between amorphous and crystalline layers and not the sample surface nearest to the electron source as this would lead to a disruption in channelling. However any hydrocarbon contamination on the surface is mobile and will lead to very thick carbonaceous deposits under the beam during imaging or acquisition of a spectrum image. The hydrocarbon molecules become polarised and are drawn to the illuminated area. So for many samples it will be necessary to bake in vacuum or to plasma clean (using either Ar/O_2 or H_2/O_2 to reduce sputtering) before introduction to the microscope column to remove the hydrocarbons.

Polishing and electro-polishing generally produce good results after cleaning. However, general samples produced by FIB are, thus far, less successful in this regard because of the thick damage layers of 10 nm thickness on each side. Low energy ion cleaning is showing promise and may improve matters. However, FIB sections may show quite large thickness variations over small lateral length scales, which may make accurate EELS spectrum imaging difficult.

Catalyst samples, particularly those that have been chemically reduced, are generally clean. But if they have to be dispersed in an organic liquid, care is needed to avoid residual hydrocarbon upon evaporation. Spent catalysts are likely to have carbonaceous or organic residues.

8.3 HAADF IMAGING

As discussed in Chapter 5, the intensity of a feature in HAADF imaging is generally a straightforward function of the atomic number and the sample thickness. This can be exploited in many circumstances to read out structures without resource to image simulation or spectroscopy. In this section we consider applications that are based primarily on direct imaging with EELS playing at most a confirmatory role.

Recently the simulation of HAADF images has been shown to provide almost perfect quantitative agreement with experiment (LeBeau and Stemmer 2008; LeBeau *et al.*, 2008) provided spatial incoherence is taken into account (see also section 5.6).

8.3.1 Imaging of Isolated Atoms

The imaging of single atoms by STEM does not require sub-Angstrom resolution *per se*. The first such experiments were performed by Crewe in the early 1970s (Crewe 1970; Wall *et al.*, 1974). He succeeded in imaging isolated U atoms on a C support film in dark-field. His instrument operated at less than 60 keV and produced a 0.3 nm probe. The atoms were visible because of the very large atomic number contrast between U (92) and C (6). As a reminder, for single atoms the intensity of the dark-field signal is proportional to Z^ζ where Z is the atomic number and ζ is considered to be between 1.5 and 2 depending on collection angles. So for a single uranium atom sitting on a support a few carbon atoms thick, the contrast is still very high and sufficient to overcome the signal to noise issue, even if the imaging probe is 0.3 nm in diameter (i.e. bigger than the nominal spatial extent of the atom).

So what do sub-Angstrom probes resulting from aberration-correction add? Because the intensity due to high atomic number atoms is more localised the signal to noise ratio is higher and much smaller differences in Z will provide sufficient contrast to be discerned in HAADF images. This is also good for imaging heavy atoms, with much improved precision in their location, on a relatively light substrate or in a relatively light matrix. However this does not identify the atom. As we shall see EELS is able to do this in favourable circumstances and is possible because of the higher current in the aberration-corrected probe.

In many circumstances it is safe to assume that the bright single atom features observed are a particular element. For example, many catalyst systems have either a highly dispersed metal on a light support or contain heavy atomic weight promoters. A particular Pt on γ-alumina catalyst was shown to consist of Pt trimers (Sohlberg et al., 2004; Pennycook et al., 2009). Careful measurement of the Pt-Pt separations combined with density functional (DFT) calculations enabled the authors to propose a model of the trimer involving an OH group. The same authors brought understanding of the means by which isolated atoms of La stabilised γ-alumina as a support (Wang et al., 2004). Very small bi-layer clusters of up to 10 Au atoms have been identified as the active phase in Au/FeO$_x$ catalysts for ambient temperature oxidation of CO (Herzig et al., 2008).

Gold is used as a catalyst for the growth of silicon nano-wires and it was feared that if gold atoms became incorporated in the growing wire the electronic properties of the wires would be adversely affected. HAADF imaging (Allen et al., 2008) showed that Au atoms decorated planar defects where they cut the surface forming parallel rows (see also Chapter 6 and Figure 6.4). There were also isolated bright features some of which were static whilst other features were mobile over time, indicating that the latter were on the surface of the Si nano-wire. A through-focal series showed that the height of the static features lay within the wire between the rows of Au atoms (Figure 8.1) confirming that the static isolated atoms were indeed in the bulk of the silicon. Multislice image simulation confirmed this behaviour.

Fischer-Tropsch catalysts for gas to liquid technologies are typically based on iron or on oxide supported cobalt. In the latter case cobalt salts are impregnated onto a support phase such as a high surface area alumina, then dried and calcined to form the spinel oxide Co_3O_4.

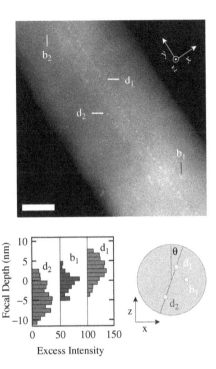

Figure 8.1 Three-dimensional localization of Au atoms. (a) Sum of seven aligned HAADF-STEM images of an intrinsic Si nanowire showing impurities trapped at a twin defect (d1,d2) and bulk impurities (b1,b2). Scale bar is 5 nm. (b) Excess intensity of Au atoms indicated in (a) plotted as a function of focal depth. The peak associated with atom b1 lies between the defect atom peaks and is therefore located within the nanowire as indicated in the schematic diagram. The dashed line extending across the nanowire represents the (111) twin defect bisecting the nanowire. (Reprinted from J. E. Allen, E. R. Hemesath, D. E. Perea, J. L. Lensch-Falk, Z.Y. Li et al., High-resolution detection of Au catalyst atoms in Si nanowires, *Nature Nanotechnology*, Vol. 3, Copyright 2008, with permission of Nature Publishing Group.) (A colour version of this figure appears in the plate section)

Promoter atoms, such as Pt, Ir, Re or Ru are added to lower the temperature at which reduction to the active metal state takes place in order to maximise the surface area and hence catalyst activity. In a useful formulation these might be about 1% (by weight) with typically 20% cobalt, the balance being the support phase. These platinum group metals are rare and expensive and minimising their loading is important to the economics of the technology. The mechanism by which they act is not understood and imaging their position and disposition is one approach to remedy this.

HAADF images of passivated samples of Co nanoparticles on alumina catalysts with each of these promoters have been published (Shannon *et al.*, 2007) and one example is given in Figure 8.2. Pt, Ir and Re all occur as isolated atoms, but the number imaged is clearly only a small fraction of those present and this observation needs to be explained.

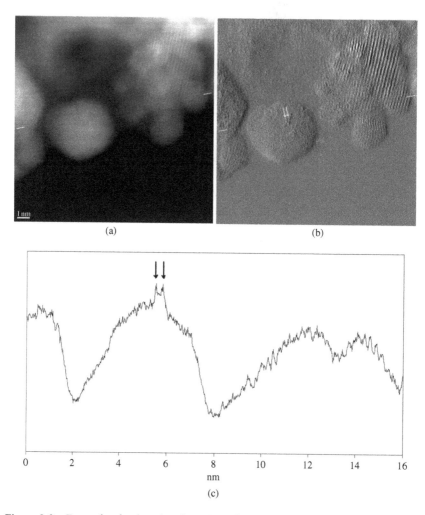

(a)   (b)

(c)

Figure 8.2   Example of reduced and passivated Ir-promoted catalyst after reduction, (a), (b) and (c). (b) is derived from the as-recorded image (a) by smoothing and application of a Kirsch East-West filter whilst (c) is a profile along the line indicated by white lines at the edge of (b). (Reprinted from M.D. Shannon, C.M. Lok and J.L. Casci, Imaging promoter atoms in Fischer–Tropsch cobalt catalysts by aberration-corrected scanning transmission electron microscopy, *J. Catalysis*, Vol. 249, p. 41–51, Copyright 2007 with permission of Elsevier.)

Consider a 3 nm cobalt metal crystallite oriented on a zone axis such that the focused electron probe is placed on a column containing n_1 atoms of atomic number Z_1 and n_2 of atomic number Z_2. A simplistic view of the HAADF intensity that results is that it is proportional to $n_1 Z_1^\zeta + n_2 Z_2^\zeta$, with $\zeta = 1.7$ (see Chapter 5). This assumes that each atom scatters independently of the others and that only a small fraction is scattered as each atom along the column is encountered. On this basis if 1 Co atom in a column of 15 were replaced by Re, the relative intensity of that column to one with 15 Co atoms would be about 1.3 compared with 1.07 if the column had an extra Co atom. A good signal to noise ratio is needed to distinguish these two possibilities for a brighter column. However this analysis is too simplistic. The electron beam experiences a focusing action by each Co atom and therefore tends to channel along the column of atoms (see sections 1.2.1 and 5.4). So if it then encounters a Re atom, a much higher signal is detected on the HAADF detector than direct illumination by the focused electron beam would have produced and the contrast is higher than predicted by the simplistic model. Conversely if a Re atom is encountered early on in the column the high scattering results in the electron flux being much less focused along the Co columns and more interstitial, so scattering from the subsequent Co atoms is reduced and the contrast is lower than predicted by the simplistic model. So heavy dopant atoms located near the exit surface of a crystal are much more visible than those near the entrance surface (Figure 8.3). This is unfortunate and ways of mitigating this need to be found to realise the full potential of HAADF STEM to image these atoms.

Other factors that reduce the observed contrast are the alumina support, that at these Co particle sizes (3–5 nm) is very difficult to avoid, and the surface passivation layer that inhibits coupling of the electron beam into the atomic columns (Mkhoyan et al., 2008).

This interpretation contrasts substantially with the work on Au atoms in Si nano-wires (Figure 8.1) where the Au atoms can be seen at both top and bottom surfaces. The atomic number of the matrix clearly has a significant role in the visibility of isolated heavy atoms. However similar complications have been reported for Sb doping of Si (Voyles et al., 2002; Voyles et al., 2003).

Of course for relatively light atoms in a matrix, EEL spectrum imaging is generally needed both to image and identify these atoms. However there is a recent example (Krivanek et al., 2010) in which all the atoms were identified by careful calibration of the intensity in the medium-angle angular dark-field (MAADF) signal. Here the substitution of atoms of

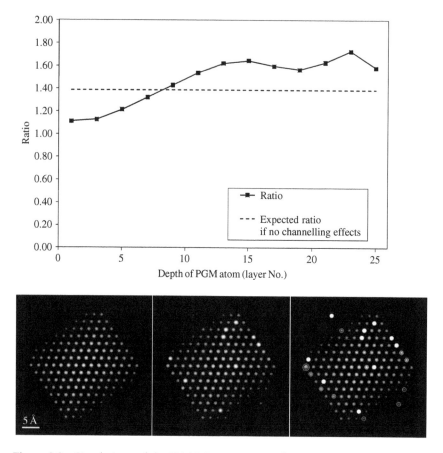

Figure 8.3   Simulations of the HAADF contrast as a function of depth of a dopant Re atom in a [110] Co slab 25 layers thick (a) and the HAADF image of an idealised 3 nm Co particle ((b)–(d)). (b) is of the undoped particle, (c) of the particle randomly doped with Re in the surface layer, and (d) indicates the positions of the dopant atoms in projection with the solid circle representing a dopant atom in the upper half of the particle (near the exit surface in the imaging process) and the open circle representing those in the lower half of the particle (near the entrance surface in the imaging process).

C and O in a monolayer of boron nitride (BN) resulting from beam damage (Figure 8.4) was identified.

Simultaneous SEM secondary electron imaging with HAADF STEM is an extremely valuable combination in studying, for example, supported metal catalysts or nanoparticles in thin tissue sections. In the former, it is useful to know which particles are on the outer surface of a porous support particle, and therefore readily available to reactant gases, and

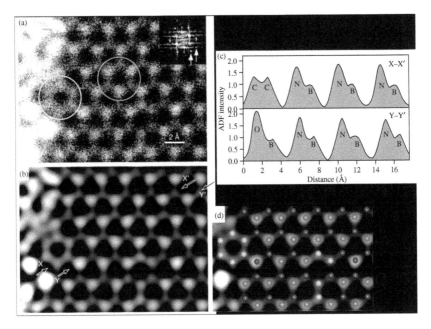

Figure 8.4 MAADF STEM image of monolayer BN. a, As recorded. b, Corrected for distortion, smoothed, and deconvolved to remove probe tail contributions to nearest neighbours. c, Line profiles showing the image intensity (normalised to equal one for a single boron atom) as a function of position in image b along X–X′ and Y–Y′. The elements giving rise to the peaks seen in the profiles are identified by their chemical symbols. Inset at top right in a shows the Fourier transform of an image area away from the thicker regions. Its two arrows point to (1120) and (2020) reflections of the hexagonal BN that correspond to recorded spacings of 1.26 and 1.09Å. Part (d) shows part of a DFT simulation of a single BN layer containing the experimentally observed substitutional impurities overlaid on the corresponding part of the experimental image. Red, B; yellow, C; green, N; blue, O. (Adapted from Krivanek *et al.*, Atom-by-Atom Structural and Chemical Analysis by Annular Dark-Field Electron Microscopy, *Nature*, Vol. 464, p571, Copyright 2010, with permission of Nature Publishing Group.) (A colour version of this figure appears in the plate section)

which are accessed via pores. The report of the imaging of single U atoms using secondary electrons (Zhu *et al.*, 2009) may lead to studies that make a valuable contribution to catalyst research (Figure 8.5).

8.3.2 Line Defects (1-D)

The reduced displacement of atomic features from their true position due to the aberration-correction in HAADF images is useful in 'reading out'

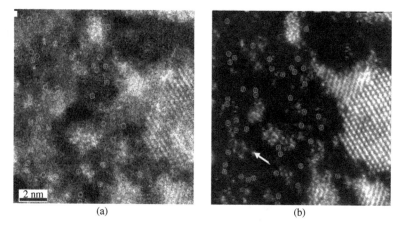

(a)   (b)

Figure 8.5 Imaging individual atoms with secondary electrons. a,b, Simultaneous acquisition of the SEM image using secondary electrons and BSEs (a), and the ADF-STEM image using transmitted electrons (b), of uranium oxide nanocrystals and individual uranium atoms (raw data). Quantised intensity analysis in the ADF image suggests the areas marked by small circles are individual isolated uranium atoms. Several spots, for instance, the one marked by an arrow, were excluded because they had double the integrated signal of the ones in the circles. The missing spots in SEM are probably the uranium atoms located on the bottom surface of the carbon support. (Reprinted from Zhu *et al.*, Imaging single atoms using secondary electrons with an aberration-corrected electron microscope, *Nature Materials*, Vol. 8, Copyright 2009 with permission of Nature Publishing Group.)

trial structures for further refinement by energy minimisation procedures. An example is provided by the study of partial dislocations in GaN semiconductors. Because of their value in providing large band-gap materials for blue and ultraviolet LEDs, there is great interest in being able to understand why these materials can tolerate much higher defect densities than other semi-conductors whilst remaining operative. Arslan, *et al.* (2005) and others have combined HAADF imaging and nitrogen K-ELNES measurements both to read out the projected atom column positions and to provide detailed spectroscopic data in an attempt to unravel the 3D structure and bonding in full-core mixed dislocations in GaN (Figure 8.6).

8.3.3 Interfaces and Extended Defects (2-D)

There are important planar interfaces in semiconductor technology; for example gate oxides in FET transistors and electrical contacts and interconnects as devices get smaller.

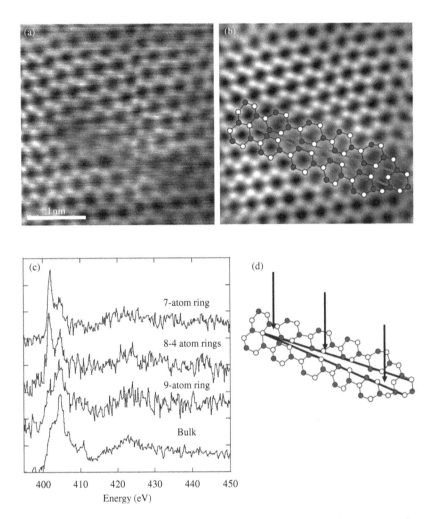

Figure 8.6 Z-contrast image of a dissociated mixed dislocation, showing a stacking fault in between the edge and screw dislocation cores (a). The ball and stick model (b) superimposed on a noise filtered copy of (a) clarifies the atomic structure and shows how the stacking fault widens in the plane of the image between the two dislocations to accommodate the strains associated with each dislocation. Other similar partial dislocations connected by stacking faults have been observed with the lengths of the stacking faults extending over 15–30 Å in length. Electron-energy loss spectra (c) from a number of positions marked in (d) along the partial dislocation, compared to bulk GaN. The fine structure of the NK-edge confirms that one dislocation has a screw character while the other has an edge character. (Adapted from I. Arslan, A. Bleloch, E.A. Stach, and N.D Browning, Atomic and Electronic Structure of Mixed and Partial Dislocations in GaN, *Phys. Rev. Lett.* Vol. 94, p. 25–504, Copyright 2005 with permission of APS.) (A colour version of this figure appears in the plate section)

In the latter case metal di-silicides are being reconsidered because of the so called fine line effect. For example $CoSi_2$ has good electrical conductivity, high thermal, chemical and mechanical stability and a large process window. It forms a perfect semiconductor/conductor junction with Si by epitaxy on the (001) plane, but the Schottky barrier between them is relatively high and hence conduction across the junction is strongly influenced by the local interface structure (Tung *et al.*, 1991) and strain due to stacking faults and dislocations involving Burgers vectors of $1/4[111]$. A great deal of effort had been expended by many microscopy groups in trying to unravel the interface structure throughout the 1980's and 1990's. A number of models were proposed, but all were limited by the inadequate resolution of TEM and STEM imaging before the advent of aberration-correction.

The imaging of this interface in the related $NiSi_2$/Si system and determination of the structure was the first major achievement of the SuperSTEM facility and was closely followed by experiments on $CoSi_2$ (Falke *et al.*, 2004; Falke *et al.*, 2005). The samples were made by co-deposition of the nickel (or cobalt) and silicon followed by a capping layer of silicon. The samples then underwent rapid thermal annealing to grow the silicides. Specimens for microscopy were thinned by tripod polishing and the surfaces cleaned by gentle ion milling.

Just one structure, known as the sevenfold R (2×1), is observed at the $NiSi_2$/Si interface whilst that and also a second, eightfold, structure are observed at $CoSi_2$/Si interfaces (note that the n-fold description relates to the coordination of Co at the interface). Two projections parallel to the interface on the Si (001) plane but at right angles to one another ($[110]$ and $[1\bar{1}0]$) are shown for both interface structures in Figure 8.7. In each case, both projections are seen along one boundary in a sample and a variety of defects are observed where these orientations change (see later). In the bulk silicide 'triplet' columns of Si, then Co, then Si are seen whilst 'doublet' columns of Si referred to as 'dumbbells' are seen in the bulk silicon, both oriented normal to the interface. In the triplet the Co atom column is clearly bright in the centre given its much higher atomic number than Si.

These pairs of projection HAADF images were sufficient to derive straightforwardly the interface structures. These structures were predicted by DFT calculations by Yu *et al.* (2001) to give the lowest interface energy difference, but this STEM study gave the first experimental verification of the sevenfold R interface. For the eightfold interface the measured position of the column of Si atoms in Figure 8.8(c) indicated

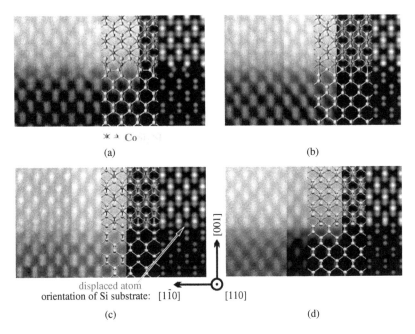

Figure 8.7 Raw HAADF image (left), averaged contrast (middle), image simulation (right), and ball and stick model (overlaid) for: the (2 × 1) reconstructed, sevenfold interface, showing two projections rotated to each other by 90° around the interface normal, (a) and (b), and the eightfold interface, (c) and (d), showing two projections again rotated. Images were taken in [110] Si substrate zone axis. Averaging over several unit cells along the interface was performed by Fourier filtering with a periodic mask. (Reprinted from M. Falke, *et al.*, Real structure of the CoSi₂/Si(001) interface studied by dedicated aberration-corrected scanning transmission electron microscopy, *App. Phys. Lett.* Vol. 86, p. 203103, Copyright 2005 with permission of American Institute of Physics.) (A colour version of this figure appears in the plate section)

that the displacement towards the silicide appears to be larger than the calculations in Yu *et al.* (2001).

Dislocations are expected at these boundaries. A typical example of the complexity of these is shown by the disclination in Figure 8.8(a). To the right and left of this image we see the two projections of the R (2 × 1) sevenfold structure of Figures 8.8(a) and 8.8(b) respectively. In the silicon there is a single column of Si atoms rather than the 'dumbbells' seen elsewhere and a micro-twin at the {111} interface facet. The atomic arrangement at the edges of the facet has yet to be unravelled. On another part of the eightfold interface a long periodicity sometimes occurs whereby Co and Si atoms swap places at every fifth

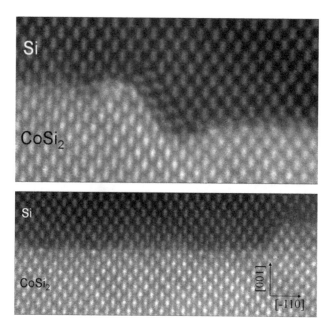

Figure 8.8 Examples of defect structures at the CoSi₂/Si (001) interface; (a) a disclination – a {111} facet revealing a 1/4 [111] type Burgers vector – and (b) ordered swapping of Co and Si atoms. (Images courtesy of M Falke and U Falke.)

position. Attempts to calculate the energy saving associated with this reorganisation have not yet been successful as the computation is very intensive.

The sharpness of interfaces between semiconductor layers is of great importance since the quality of these structures influences the electronic and optical properties of modern high speed electronic devices. A careful study of the contrast in HAADF images of AlAs/GaAs multilayers (Robb *et al.*, 2008, 2010) has shown that it is possible to measure the local interface sharpness. They use column intensity ratio mapping to reveal the compositional distribution across the whole HAADF image and this allows a statistical analysis and an estimation of errors. In this method the background intensity is removed by Fourier filtering and the intensity ratio of the Group III column (Ga, Al or mixture of the two) to the Group V column (As) calculated for every III–V dumbbell in the [110] image. (see Figure 8.9)

These examples indicate the power of HAADF to provide directly interpretable images in many circumstances. However caution is required when heavy dopant atoms are involved.

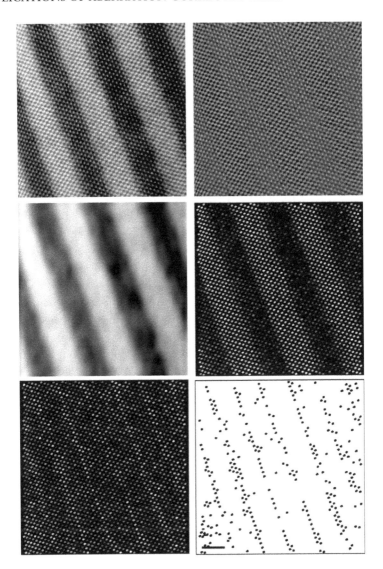

Figure 8.9   (a) HAADF image of several repeats of a 9 ML (monolayer) AlAs/9 ML GaAs superlattice in which the GaAs layers appear as the most intense due to the underlying background signal. (b) The background removed image of (a). (c) The removed background signal. (d) The dumbbell column ratio map. Bright dumbbells have high column ratios and are more GaAs-like. Dark dumbbells have low column ratios and are more AlAs-like. (e) The column ratio map of (d) re-plotted as a standard deviation departure map. (f) The column ratio map re-plotted to highlight the positions of the dumbbells that have ratios intermediate between AlAs and GaAs. The scale bar is 2 nm. (Adapted from P.D. Robb and A.J. Craven, Column ratio mapping: A processing technique for atomic resolution high-angle annular dark-field (HAADF) images, *Ultramicroscopy*, Vol. 109, p. 61–69, Copyright 2008 with permission of Elsevier.)

8.3.4 Detailed Particle Structures (3-D)

Aberration-corrected STEM instruments do have a valuable role to play in biological and medical applications, particularly where inorganic materials interface with tissue. One example is the study of the mechanism and pathology of iron storage in plants and animals which occurs via the iron storage protein ferritin which is composed of a 6 nm diameter mineral core surrounded by a protein shell. In trying to understand the mechanism for the biomineralisation and release of iron in ferritin molecules in fixed cellular tissue sections (in this case the liver), the structure and architecture of the ferritin mineral core has been unravelled, the former by atomic resolution HAADF imaging and the latter by 'single particle' analysis (Figure 8.10, Pan *et al.*, 2006–2009). The structure of the mineral core is found to be based on a phosphorous-doped two (XRD) line ferrihydrite. Single particle analysis is a statistical method which assumes all particles are the same but are oriented in different projections, the method being based on the classification of similar projections (Figure 8.10b) from an otherwise random set of projections and creates a reconstruction that is consistent with the classes of projections. In images such as Figure 8.10a, only certain ferritin mineral cores were selected on the basis of their integrated HAADF intensities, which were shown to be directly linked to iron loading via absolute EELS quantification via spectrum imaging (see sections 7.5.1 and 7.7). Hence a subset of

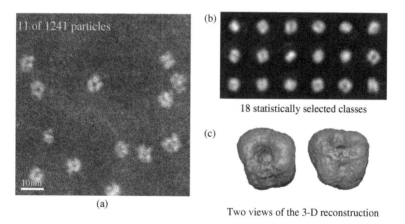

(b)

18 statistically selected classes

(c)

Two views of the 3-D reconstruction

(a)

Figure 8.10 (a) An example of one of 133 HAADF images containing multiple ferritin cores, used for the reconstruction. (b) Core images all bandpass-filtered, aligned and classified using the EMAN program into the 18 projection classes. (c) Two views of the 3-D reconstruction. (Figure courtesy of Y Pan and K Sader.)

particles which had similar iron loadings were chosen for single particle analysis. The resultant reconstruction (Figure 8.10c) and hence mineral core morphology is that of linked sub-grains at the vertices of a cube (though not all vertices are occupied on average) together with a void at the centre. This sub-unit morphology appears to be templated by the symmetry of the surrounding protein shell.

Some data on the surface and interface chemistry in the mineral core has also been gleaned by EELS analysis at the edge of particles. This is extremely difficult because of electron beam damage to the radiation-sensitive ferrihydrite. Many analyses with very short acquisition/dwell times averaged over many particles are required and 'smart acquisition' methods are being developed to do this under computer control (see sections 6.4.5 and 7.7).

On occasion the linear relationship between HAADF intensity and projected thickness can be turned into the local z components of the upper and lower surfaces of a particle and the shape and facets identified from a single projection. This is true even in the case of $NiSi_2$ precipitates in a Si matrix (Figure 8.11). The projection is along a zone-axis orientation and *a priori* knowledge of interface planes in the system is needed in order to extract the detailed particle shape.

Another example of this involves Au_{309} particles which are difficult to image as they are extremely mobile on the support: successive scans

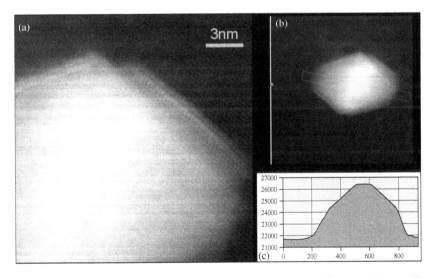

Figure 8.11 HAADF images (a), (b) of a $NiSi_2$ precipitate in a Si matrix. The profile from the averaged linescan in (c) indicates the presence of facets on the precipitate.

reveal different orientations of the particle and isolated atoms on the C support that have diffused away from the particles. Li *et al.* (2008) succeeded in imaging these by HAADF. In the lower part of Figure 8.12, the profile from the central atom column of the particle to one of the corners is in good agreement with the Ino-decahedral model oriented parallel to the five-fold axis except for the last column of atoms. This column appears to have too few Au atoms, but a shoulder next to this column (arrowed) suggests that a Au atom has left the particle.

More generally tomography is an important method for revealing the 3-D structure of heterogeneous materials. Conventional electron tomography employs 100–200 projections with a 1° separation and a back-projection reconstruction. The voxel resolution is currently about 1 nm for materials science applications and 2 nm for cryo-tomography of biological systems. However a number of authors have recognised the possibilities of achieving much better resolution with a small number of projections, so called Discrete Tomography, using the linear correspondence between HAADF intensity and thickness for sub 10 nm particles (Bals *et al.*, 2007; Jinschek *et al.*, 2008). An example of this is determining the facets on a 6 nm Pt particle using just 5 projections is given in Figure 8.13 (Jiang *et al.*, 2011). Pt particles of this size and smaller are important in many catalyst applications including emerging fuel-cell technologies. The proportion of each surface plane and edge is of interest. This method unlike other proposed methods for discrete tomography does not require zone-axis projections, just projections with accurately known orientation relationships. In order to apply this method to a supported catalyst system, a large number of particles would need to be reconstructed and these would be at different heights in or on a fragment of support material (for example carbon). A through focal series will be required at each tilt to compensate for the much reduced depth of focus of aberration-corrected instruments. So in practice the number of images will be similar to conventional tomography but will be easier and quicker to obtain.

A common type of sample for aberration-corrected STEM is core-shell nanoparticles. The detailed 3-D structure is of interest but difficult to determine. HAADF imaging and EELS mapping go some way to providing this. It is not uncommon for such particles to have a different structure from that designed by the synthetic chemist: the core and shell may be inverted, or the shell may be particulate and not fully enclose the core, or the shell may be present as a separate phase. Two methods that are being developed will play a significant role in the characterisation of these particles. Mendis and Craven (2010) describe a method for

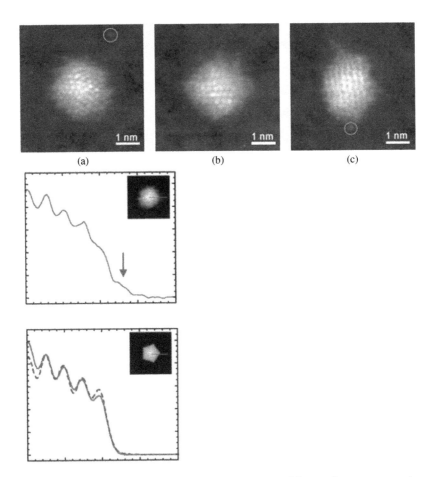

(a)   (b)   (c)

Figure 8.12   High-resolution HAADF-STEM images of Au_{309} clusters on a carbon film. Typical images show various outline shapes, that is, cluster projections: pentagon (a), square (b) and hexagon (c). The intensity variation within the clusters clearly demonstrates atomic column resolution (reproduced from Li *et al.*, 2008). The lower part of the figure shows the three dimensional atomic structure derived from Figure 8.12a (mirrored across a vertical line): Experimental intensity line profile taken from the central atom column of the cluster to one of the corners (indicated in inset with red line. C, Simulated HAADF-STEM image (inset), obtained with a simple kinematical approach, of an Au_{309} cluster with Ino-decahedral geometry. An intensity profile (solid curve) across one ridge (indicated in inset by a grey line) is compared with the result from a full dynamical multislice calculation (dashed line). (Reprinted from Z.Y. Li, N.P. Young, M.D. Vece, S. Palomba, R.E. Palmer, A.L. Bleloch, B.C. Curley, R.L. Johnston, J. Jiang and J. Yuan, *Nature, 451*, 06470 (2008).)

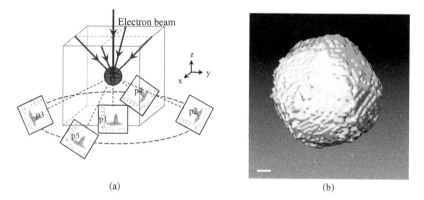

(a)   (b)

Figure 8.13 (a) Schematic showing the five projections needed for the reconstruction and (b) one view of the reconstruction of the 6 nm Pt particle (scale bar 1 nm). (A colour version of this figure appears in the plate section)

extracting separate core and shell spectra from core-shell particles with varying core to shell volume fractions; whilst Hondow *et al.* (2010a) describe the use of Diffraction Imaging (i.e. recording a nanodiffraction pattern at each STEM probe position) to map the structure of the core and sub-grains in the shell of core-shell particles.

8.3.5 Low-loss EELS

In general there is concern to establish what changes, if any, result in our bodies from exposure to nanoparticles. HAADF is a particularly useful method to image these nanoparticles since most are inorganic and have a higher mean atomic number than biological tissue (Hondow, (2010b)). Unlike high atomic number inorganic nanoparticles, single-wall carbon nanotubes (SWCNT), which are being considered as contrast agents for medical imaging and for the delivery of therapeutically active molecules to target cells, are particularly challenging to locate in tissue because of the similarity in atomic number. The slight difference in plasmon peak energy between the nanotubes and the tissue has been exploited to locate them at a relatively low magnification using EELS spectrum imaging prior to imaging carefully by BF STEM (Figure 8.14) (Porter *et al.*, 2007, Gass *et al.*, 2010). This contributed to a significant study that demonstrated that individual acid-treated SWNTs found inside lysosomes and also the cytoplasm caused no significant changes in cell viability or structure even after 4 days of exposure (Porter *et al.*, 2009).

Graphene and other 2D crystals are being heavily studied because of the substantial change in properties between bulk 3D crystals and single

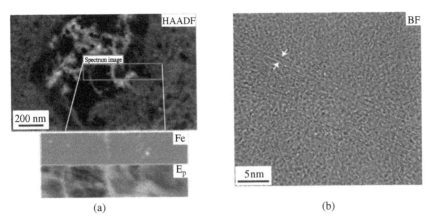

(a)   (b)

Figure 8.14 Intracellular distribution of SWNTs in unstained sections. a, HAADF-STEM image of SWNTs within a lysosome (two days' exposure). Box shows low-loss EELS image: Fe map (white); plasmon map (Ep) ranges from 17.5 eV (dark) to 21.5 eV (light). b, High-resolution bright-field (BF) image of SWNTs. Arrows indicate the diameter of SWNTs. (Reprinted from A.E. Porter, M. Gass, K. Muller, J. Skepper, P.A. Midgley and M. Well, Direct imaging of single-walled carbon nanotubes in cells, *Nature Nanotechnology*, Vol. 2, p. 713–717, Copyright 2007, with permission of Nature Publishing Group.)

sheets. This is another example were low-loss EELS is particularly useful in identifying which regions are single layers – most such materials not yet being produced as completely monolayer crystals. Gass *et al.* (2008) showed that there is a shift to significantly lower energy of the plasmon peak as the number of layers is reduced. Bulk graphite has a plasmon energy of 27 eV (beam normal to sheets) whereas five layers have a peak at about 17 eV and a single layer of graphene has a (surface) plasmon peak at about 15 eV (Figure 8.15). Studies of defects and their movement that modify the properties of these materials can then be undertaken on areas that have been established to be monolayers or otherwise.

8.3.6 Core-loss EELS and Atomic-scale Spectroscopic Imaging

The first experiment designed to demonstrate the combination of imaging and chemical identification of a single heavy dopant atom was due to Varela *et al.*, 2004 (Figure 8.16). Occasional random lanthanum substitution for calcium in a calcium titanate crystal was imaged in HAADF and the characteristic 'white lines' due to the EELS La $M_{4,5}$ edge at 830 eV loss were recorded with the electron probe focused on

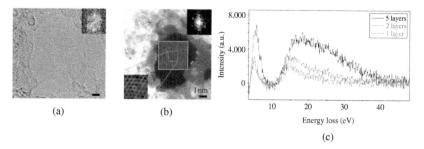

(a)   (b)   (c)

Figure 8.15   High-resolution images of mono-layer graphene. a, b, Bright-field (a) and HAADF (b) images of the monolayer, showing a clean patch of graphene surrounded by a mono-atomic surface layer; individual contaminant atoms of higher atomic number can be seen in b. The inset FFT clearly shows the lattice in the HAADF image and, by applying a bandpass filter, the atomic structure is apparent. c, an EELS spectrum from the monolayer is compared with that from a bi-layer and bulk graphene (graphite); Scale bars in a and b, 1 nm. (Adapted from Gass *et al.*, Free-Standing Graphene At Atomic Resolution, *Nature Nanotechnology*, Vol. 3, p676–681, Copyright 2008, with permission from Nature Publishing Group.) (A colour version of this figure appears in the plate section)

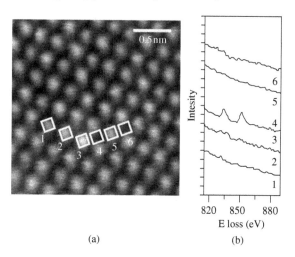

(a)   (b)

Figure 8.16   (a) Z-contrast image with (b) EELS traces showing spectroscopic identification of a single La atom at atomic spatial resolution, with the same beam used for imaging. The $M_{4,5}$ lines of La are seen strongly in spectrum 3 obtained from the bright column at 2×10^7 magnification and a total collection time of 30s. Other spectra from neighbouring columns show much reduced or undetectable La signal. These spectra were obtained with collection times of 20 s, and are shown normalized to the pre-edge intensity and displaced vertically for clarity. (Reprinted from Varela, M., Findlay, S.D., Lupini, A., Christen, Borisevich, A.Y., Delby, N., Krivanek, O.L., Nellist, P., Oxley, M.P., Allen, L.J. and Pennycook, S.J. (2004) Spectroscopic Imaging of Single Atoms Within a Bulk Solid, *Phys. Rev. Lett.*, 92, 095502 with permission of the American Physical Society.)

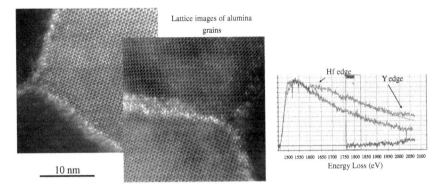

Figure 8.17 HAADF images of alumina grains and associated boundaries (a), (b). In (c) five summed EELS spectra from boundaries showing Hf and Y edges (upper trace) are compared with the spectrum from neighbouring alumina (middle trace). (Figure courtesy of D Ram and G J Tatlock.)

the column containing one dopant atom and whilst they were almost absent when it was focused on adjacent atom columns.

Other examples of this procedure include imaging and identification of a single dopant atoms in few layer graphene and ion-implanted MWCNT samples (Bangert *et al.*, 2010) as well as Hf atoms segregated to grain boundaries in alpha alumina (Figure 8.17, Tatlock *et al.*, 2009). In the latter case however the EELS spectrum is the sum of those obtained from 5 bright atoms in the HAADF image in order to improve the signal to noise ratio, but it is recorded at a much higher energy loss of around 1600 eV.

It would appear that the prospects are good for further exploitation of this combination of imaging and analysis at the atomic scale of high atomic number species in materials of technological importance. Figure 8.17 is one such example. Metal alloys for high temperature applications have a thermally grown alumina scale to act as a barrier to further oxidation. In growing this scale, reactive elements such as Hf and Y are added. Their effects are to slow down the growth of oxide (having allowed it to form quickly in the first instance), improve the adherence of the scale and improve the plasticity of the scale. In use the rate of oxide growth eventually rises again. Chen *et al.* (2009) and Tatlock *et al.* (2009) have shown how the Hf and Y atoms are initially distributed as isolated atoms and are not clustered in the grain boundaries and then by a process of Ostwald ripening migrate to the largest precipitates at triple points.

As discussed in Chapter 7, with the high current and (sub) 0.1 nm probes in an aberration-corrected STEM the prospect of forming images at the atomic scale by EELS spectrum imaging using electrons characteristic of at least some of the atoms in the sample is a realisable goal. To do so the sample needs to be reasonably beam stable and have edges preferably in the range 300 to 1000 eV loss. The major stumbling block has been the signal to noise ratio in the recorded spectrum image. Bosman et al. (2007) developed a method of oversampling to correct for detector irregularities and significantly improve the signal to noise ratio. A proof of principle example showing the location of Mn and O in a mixed perovskite is given in Figure 8.18. Here, the experimental data is compared with Bloch wave simulations using effective inelastic scattering potentials calculated using the method described by Allen et al. (2003). The zigzag pattern of the oxygen columns in the simulations for the [110] orientation is just recognisable in the experimental data and is caused by the presence of tilted MnO_6 octahedra.

Almost simultaneously, papers on atomic column EELS were submitted by Kimoto et al. (2007) and Muller et al. (2008). In the latter case new information is also obtained. A $La_{0.7}Sr_{0.3}MnO_3/SrTiO_3$ multilayer was examined and Mn-Ti intermixing on the B-site sub-lattice observed at the interface between the two layers (Figure 8.19).

Whilst EDX analysis and mapping is commonplace on FEG-TEM/STEM instruments, EELS has been preferred on the dedicated aberration-corrected STEM instruments. EELS is generally a more sensitive analytical method but there are difficulties associated with high atomic number elements where the available edges are either at very high energy loss (>2 keV) and hence very weak, or at low energies (<100 eV) and therefore the signal is delocalised and superimposed on a very steep background (see section 7.8). So there are circumstances where EDX is essential. As discussed in section 7.8, Watanabe et al. (2010a and 2010b) have demonstrated that the increased current for a given probe size in aberration-corrected STEM when combined with multivariate statistical analysis makes atomic column resolution EDX mapping of gallium arsenide distinctly possible (Figure 8.20).

With the arrival of SDD detectors (section 7.2.1) there has been a resurgence of interest in EDX in aberration-corrected STEM. There is no liquid nitrogen boil-off that might affect the mechanical stability of the STEM and there is the prospect of much larger solid angles of collection. The standard solid angle has been around 0.13 steradian (1% of the emitted characteristic X-rays) for three decades with a

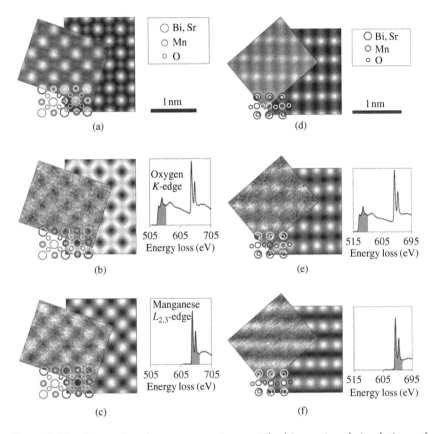

Figure 8.18 Comparison between experiments (tilted images) and simulations of $Bi_{0.5}Sr_{0.5}MnO_3$ oriented along the [001] zone axis (a–c) and along the [110] axis (d–f). (a) Z-contrast, (b) oxygen K edge and (c) manganese $L_{2,3}$ edge STEM images in [001] orientation. The atomic structure is indicated. The EEL maps were generated by integrating the EEL spectra over a 30 eV window above the respective ionization threshold. This is shown in gray shading on the spectra to the right, which are summations of all the spectra in the adjacent EELS maps. The Z-contrast image is the summation of three cross-correlated single images, taken from the same area, with the same experimental settings as the EELS maps. The simulations assume a 330Å thick sample. (d) Z-contrast, (e) oxygen K edge and (f) manganese $L_{2,3}$ edge STEM images along [110] zone axis. The simulations assume a 120Å thick sample. (Reprinted from M. Bosman, V.J. Keast, J.L. Garcia-Munoz, A.J.D. Alfonso, S.D. Finlay, and L.J. Allen, Two-Dimensional Mapping of Chemical Information at Atomic Resolution, *Phys. Rev. Lett.* Vol. 99, p. 86–102, Copyright 2007, with permission from American Physical Society.)

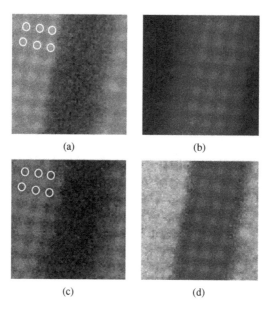

Figure 8.19   Spectroscopic imaging of a $La_{0.7}Sr_{0.3}MnO_3/SrTiO_3$ multilayer, show-ing the different chemical sublattices. (a) La M edge; (b) Ti L edge; (c) Mn L edge; (d) red-green-blue false colour image obtained by combining the rescaled Mn, La, and Ti images. Each of the primary colour maps is rescaled to include all data points within two standard deviations of the image mean. The white circles indicate the position of the La columns, showing that the Mn lattice is offset. Live acquisition time for the 64×64 spectrum image was ~ 30 s; field of view, 3.1 nm. (Reprinted from Muller, D.A., Kourkourtis, L.F., Murfitt, M., Song, J.H., Hwang, H.Y., Silcox, J., Delby, N. and Krivanek, O.L. (2008) Atomic-Scale Chemical Imaging of Composition and Bonding by Aberration-Corrected Microscopy, Science, 319, p1073–1076 with permission of the American Association for the Advancement of Science.) (A colour version of this figure appears in the plate section)

few exceptions, but at the time of writing, recent developments from manufacturers offer solid angles of 0.9 steradian and more.

8.4 CONCLUSIONS

Aberration-corrected analytical STEM has made possible real-space crystallography and directly interpretable imaging of defect structures. In doing so it has revolutionised the characterisation of materials for nanotechnology. It has applications ranging from catalysis, nanoparticles and quantum dots through semiconductors and ceramics to life-sciences.

After a period when many publications highlighted instrument perfor-mance or gave proof of principle experiments of new capability afforded

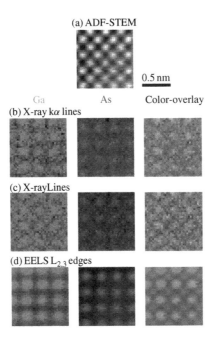

Figure 8.20 (a) HAADF-STEM image of the [001]-projected GaAs simultaneously recorded during SI acquisition; a set of a Ga maps, and As maps and a color-overlay map of K X-ray lines (b), L X-ray lines (c) and EELS $L_{2,3}$ edges (d). (Reprinted from Watanabe, M., Kanno, M. and Okunishi, E. (2010a) Atomic-Resolution Elemental Mapping by EELS and XEDS in Aberration Corrected STEM, *Jeol. News*, 45(1), p 8–15, with permission of Watanabe, M.) (A colour version of this figure appears in the plate section)

by aberration-corrected STEM instruments, real materials applications are now starting to dominate the literature and will increasingly do so in future years.

REFERENCES

Allen, J.E., Hemesath, E.R., Perea, D.E., Lensch-Falk, J.L., Li, Z.Y., Yin, F., Gass, M.H., Wang, P., Bleloch, A.L., Palmer, R.E. and Lauhon, L.J. (2008) High-Resolution Detection of Au Catalyst Atoms in Si Nanowires, *Nature Nanotech*, 3, 168–173.

Allen, L.J., Findlay, S.D. Oxley, M.P., and Rossouw, C.J. (2003) Lattice-Resolution Contrast From a Focused Coherent Electron Probe. Part I, *Ultramicroscopy*, 96, 47.

Arslan, I., Bleloch, A., Stach, E.A., and Browning, N.D. (2005) Atomic and Electronic Structure of Mixed and Partial Dislocations in GaN, *Phys. Rev. Lett.*, 94, 025504.

Bals, S., Batenburg, K.J., Verbeeck, J., Sijbers, J. and Van Tendeloo, G. (2007) Quantitative Three-Dimensional Reconstruction of Catalyst Particles for Bamboo-like Carbon Nanotubes, *Nano Letters*, 7, 3669–3674.

Bangert, U., Bleloch, A., Gass, M., Seepujak, A. and van den Berg, J. (2010) Doping of Few-Layered Graphene and Carbon Nanotubes Using Ion Implantation, *Phys. Rev. B.*, 81, 245423.

Bosman, M., Keast, V.J., Garcia-Munoz, J.L., Alfonso, A.J.D., Finlay, S.D. and Allen, L.J. (2007) Two-Dimensional Mapping of Chemical Information at Atomic Resolution, *Phys. Rev. Lett.*, 99, 086102.

Chen, C.-L., Wang, P. and Tatlock, G.J. (2009) Phase Transformations in Yttrium-Aluminium Oxides in Friction Stir Welded and Recrystallised PM2000 Alloys, *Mater. High. Temp.*, 26, 299–303.

Crewe, A.V. (1970) Visibility of Single Atoms, *Science*, 168, 1338–1340.

Falke, M., Falke, U., Bleloch, A., Teichert, S., Beddies, G. and Hinneberg, H.-J. (2005) Real Structure of the $CoSi_2/Si$ (001) Interface Studied by Dedicated Aberration Corrected Scanning Transmission Electron Microscopy, *App. Phys. Lett.*, 86, 203103.

Falke, U., Bleloch, A., Falke, M. and Teichert, S. (2004) Atomic Structure of a (2×1) Reconstructed $NiSi_2/Si(001)$ Interface, *Phys. Rev. Lett.*, 92, 116103.

Gass, M.H., Bangert, U., Bleloch, A.L., Wang, P., Nair, R.R. and Geim, A.K. (2008) Free-Standing Graphene At Atomic Resolution, *Nature Nanotechnology*, 3, 676–681.

Gass, M.H., Porter, M.H., Bendall, J.S., Skepper, J.N., Midgley, P.A., Muller, K. and Welland, M. (2010) Cs Corrected STEM EELS: Analysing Beam Sensitive Carbon Nanomaterials in Cellular Structures, *Ultramicroscopy*, 110, 946–951.

Herzig, A.A., Kiely, C.J., Carley, A.F., Landon, P. and Hutchings, G.J. (2008) Identification of Active Gold Nanoclusters on Iron Oxide Supports for CO Oxidation, *Science*, 321 1331–1335.

Hondow, N.S., Chou, Y.-H., Sader, K., Douthwaite, R.E. and Brydson, R. (2010a) Electron Microscopy of Cocatalyst Nanostructures on Semiconductor Photocatalysts, *J. PhysChem C.*, 114, 22758–22762.

Hondow, N., Harrington, J., Brydson, R., Doak, S.H., Singh, N., Manshian, B. and Brown, A.P. (2010b) STEM Mode in the SEM: A Practical Tool for Nanotoxicology, *Nanotoxicology* doi:10.3109/17435390.2010.535622.

Jiang, L., Wang, P., Goodhew, P.J., Shannon, M.D. and Bleloch (in preparation 2011).

Jinschek, J.R., Batenburg, K.J., Calderon, H.A., Kilaas, R., Radmilovic, V. and Kisielowski, C. (2008) 3-D Reconstruction of the Atomic Positions In a Simulated Gold Nanocrystal Based On Discrete Tomography: Prospects of Atomic Resolution Electron Tomography, *Ultramicroscopy* 108, 589–604.

Kimoto, K., Asaka, T., Nagai, T., Saito, M., Matsui, Y. and Isizuka, K. (2007) Element-Selective Imaging of Atomic Columns in a Crystal Using STEM and EELS, *Nature*, 450, 702–704.

Krivanek, O.L., Chisolm, M.F., Nicolosi, V., Pennycook, T.J., Corbin, G.J, Delby, N., Murfitt, M.F., Own, C.S., Szilagy, Z.S., Oxley, M.P., Pantelides, S.T. and Pennycoook, S.J. (2010) Atom-by-Atom Structural and Chemical Analysis by Annular Dark-Field Electron Microscopy, *Nature*, 464, 571.

LeBeau, J.M., Findlay, S.D., Allen, L.J. and Stemmer, S. (2008) Quantitative Atomic Resolution Scanning Transmission Electron Microscopy, *Phys. Rev. Lett.*, 100, 206101.

LeBeau, J.M. and Stemmer, S. (2008) Experimental Quantification of Annular Dark-Field Images in Scanning Transmission Electron Microscopy, *Ultramicroscopy*, 108, 1653–1658.

Li, Z.Y., Young, N.P., Vece, M.D., Palomba, S., Palmer, R.E., Bleloch, A.L., Curley, B.C., Johnston, R.L., Jiang, J. and Yuan, J. (2008) Three-Dimensional Atomic-Scale Structure of Size-Selected Gold Nanoclusters, *Nature Nature*, 46–48.

Mendis B.G. and Craven, A.J. (2010) Characterising the Surface and Interior Chemistry of Core–Shell Nanoparticles Using Scanning Transmission Electron Microscopy, *Ultramicroscopy*, 111, 212–216.

Mkhoyan, K.A., Maccagnano-Zacher, S.E., Kirkland, E.J. and Silcox, J. (2008) Effects of Amorphous Layers On ADF-STEM Imaging, *Ultramicroscopy*, 108, 79–803.

Muller, D.A., Kourkoutis, L.F., Murfitt, M., Song, J.H., Hwang, H.Y., Silcox, J., Delby, N. and Krivanek, O.L. (2008) Atomic-Scale Chemical Imaging of Composition and Bonding by Aberration-Corrected Microscopy, *Science*, 319 1073–1076.

Pan, Y., Brown, A., Brydson, R., Warley, A. and Powell, J. (2006) Electron Beam Damage Studies of Synthetic 6-Line Ferrihydrite and Ferritin Molecule Cores Within a Human Liver Biopsy, *Micron*, 37, 403–411.

Pan, Y.-H., Brown, A., Sader, K., Brydson, R., Gass, M. and Bleloch, A. (2008) Quantification of Absolute Iron Content In Mineral Cores of Cytosolic Ferritin Molecules In Human Liver, *Mater. Sci. Technol.*, 24, 689–694.

Pan, Y.-H., Sader, K., Powell, J., Bleloch, A., Gass, M., Trinick, J., Warley, A., Li, A., Brydson, R. and Brown, A. (2009) New Evidence For a Subunit Structure Revealed By Single Particle Analysis of HAADF-STEM Images, *J. Struct. Biol.*, 166, 22–31.

Pennycook, S.J., Chisolm, M.F., Lupini, A.R., Varela, M., Borisevich, A.Y., Oxley, M.P., Luo, W.D., van Benthem, K., Oh, S.-H., Sales, D.L., Molina, S.I., Garcia-Barriocanal, J., Leon, C., Sanatmaria, J., Rashkeev, S.N. and Pantilides, S.T. (2009) Aberration-Corrected Scanning Transmission Electron Microscopy: From Atomic Imaging and Analysis to Solving Energy Problems, *Phil. Trans. Roy. Soc. A.*, 367, 3709–3733.

Porter, A.E., Gass, M., Muller, K., Skepper, J., Midgley, P.A. and Welland, M. (2007) Direct Imaging of Single-Walled Carbon Nanotubes In Cells, *Nature Nanotechnology*, 2, 713–717.

Porter, A.E., Bendall, J.S, Muller, K., Goode, A., Skepper, J.N., Midgley, P.A., Gass, M.H. and Welland, M. (2009) Uptake of Noncytotoxic Acid-Treated Single-Walled Carbon Nanotubes into the Cytoplasm of Human Macrophage Cell, *ACS Nano.*, 3 1485–1492.

Robb, P.D. and Craven, A.J. (2008) Column Ratio Mapping: A Processing Technique For Atomic Resolution High-Angle Annular Dark-Field (HAADF) Images, *Ultramicroscopy*, 109, 61–69.

Robb, P.D., Finnie, M., Longo, P. and Craven, A.J. Experimental Evaluation of Interfaces Using Atomic-Resolution High Angle Annular Dark Field (HAADF) Imaging, *Ultramicroscopy* (submitted 2010).

Shannon, M.D., Lok, C.M. and Casci, J.L. (2007) Imaging Promoter Atoms In Fischer–Tropsch Cobalt Catalysts By Aberration-Corrected Scanning Transmission Electron Microscopy, *J. Catalysis*, 249, 41–51.

Sohlberg, K., Rashkeev, S., Borisevich, A.Y., Pennycook, S.J. and Pantelides, S.T. (2004) Origin of Anomalous Pt–Pt Distances in the Pt/Alumina Catalytic System, *Chemphyschem*, 5, 1893–1897.

Tatlock, G.J., Ram, D. and Wang, P. (2009) High Spatial Resolution Imaging of the Segregation of Reactive Elements To Oxide Grain Boundaries In Alumina Scales, *Mater. High. Temp.*, 26, 293–298.

Tung, R.T., Levi, A.F.J., Sullivan, J.P. and Schrey, F. (1991) Schottky-Barrier Inhomogeneity At Epitaxial $NiSi_2$ Interfaces On Si(100), *Phys. Rev. Lett.*, 66, 72.

Varela, M., Findlay, S.D., Lupini, A., Christen, Borisevich, A.Y., Delby, N., Krivanek, O.L., Nellist, P., Oxley, M.P., Allen, L.J. and Pennycook, S.J. (2004) Spectroscopic Imaging of Single Atoms Within a Bulk Solid, *Phys. Rev. Lett.*, 92, 095502.

Voyles, P.M., Grazul, J.L. and Muller, D.A. (2003) Imaging Individual Atoms Inside Crystals With ADF-STEM, *Ultramicroscopy*, 96, 251.

Voyles, P.M., Muller, D.A., Grazul, J.L., Citrin, P.H. and Gossman, H.-J.L. (2002) Atomic-Scale Imaging Of Individual Dopant Atoms and Clusters In Highly n-Type Bulk Si, *Nature*, 416, 826–829.

Wall, J., Langmore, J., Isaacson, M. and Crewe, A.V. (1974) Scanning Transmission Electron Microscopy at High Resolution, *Proc Nat. Acad. Sci. USA*, 71, 1.

Wang, S., Borisevich, A.Y., Rashkeev, S.N., Glazoff, M.V., Sohlberg, K., Pennycook, S.J. and Pantelides, S.K. (2004) Dopants Adsorbed As Single Atoms Prevent Degradation of Catalysts, *Nature Materials*, 3, 143–146.

Watanabe, M., Kanno, M. and Okunishi, E. (2010a) Atomic-Resolution Elemental Mapping by EELS and XEDS in Aberration Corrected STEM, *Jeol. News*, 45, Issue 1, 8–15.

Watanabe, M., Okunishi, M. and Aoki, T. (2010b) Atomic-Level Chemical Analysis by EELS and XEDS in Aberration-Corrected Scanning Transmission Electron Microscopy, *Microsc. Microanal.*, 16, 66–67.

Yu, B.D., Miyamamoto, Y., Sugino, O., Sakai, A., Sasaki, T. and Ohno, T. (2001) Structural and Electronic Properties Of Metal-Silicide/Silicon Interfaces: A first-Principles Study, *J. Vac. Sci. Technol. B.*, 19, 1180.

Zhu, Y., Inada, H., Nakamura, K. and Wall, J. (2009), Imaging Single Atoms Using Secondary Electrons With An Aberration-Corrected Electron Microscope, *Nature Materials*, 8, 808–812.

9

Aberration-Corrected Imaging in CTEM

Sarah J. Haigh[1,2] and Angus I. Kirkland[1]

[1] *Department of Materials, University of Oxford, Oxford, UK*
[2] *University of Manchester, Materials Science Centre, Manchester, UK*

9.1 INTRODUCTION

Although the electron optical configuration of the Conventional Transmission Electron Microscope (CTEM) is quite different to that of the Scanning Transmission Electron Microscope (STEM) (as discussed in Chapters 1–3), both are well established tools for atomic resolution structural characterisation. The information provided by these two techniques is complementary and they are applicable to a wide range of materials (for reviews see Smith (1997) and Kirkland, Chang and Hutchison (2007)). Consequently, many CTEM instruments are fitted with beam scan coils and are thus able to provide both CTEM and STEM imaging modes. Despite the CTEM being relatively well established before STEM was invented, aberration-correction for the broad beam CTEM imaging mode took longer to evolve than aberration-correction in STEM (see section 9.2). At the time of writing there are still many more instruments capable of aberration-corrected STEM imaging than there are aberration-corrected CTEMs, although comparable resolutions are achievable using both imaging modes. In addition,

Aberration-Corrected Analytical Transmission Electron Microscopy, First Edition.
Edited by Rik Brydson.
© 2011 John Wiley & Sons, Ltd. Published 2011 by John Wiley & Sons, Ltd.

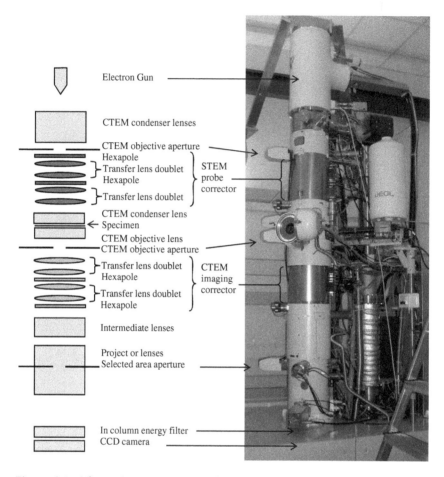

Figure 9.1 The JEOL 2200MCO column fitted with both imaging (CTEM) and STEM (probe forming) aberration-corrector elements. Labels refer to CTEM imaging.

there are a minority of microscopes fitted with both probe-forming and image-side aberration-correctors and which are hence capable of both aberration-corrected STEM and CTEM imaging (Figure 9.1). In this chapter we present an overview of aberration-corrected CTEM and demonstrate how this imaging mode provides atomic resolution structural information that is complementary to that retrieved in STEM. As in Chapter 5, some of the detailed mathematics is presented separately in boxes, with a more intuitive physical explanation presented in the text.

9.2 OPTICS AND INSTRUMENTATION FOR ABERRATION-CORRECTED CTEM

9.2.1 Aberration-Correctors

As discussed in section 4.3.1, Scherzer published configurations for correcting the inherent spherical aberration in round electromagnetic objective lenses by the use of non-round optical elements in 1947. Several groups subsequently developed practical correctors based on these proposals but were unable to demonstrate any resolution improvement. Spherical aberration-correction *with* a directly interpretable resolution improvement from 0.24 nm to 0.12 nm was finally achieved in 1997 (Haider *et al.*, 1998) with a corrector based on a design by Rose.

In this corrector (Figure 9.1 and section 4.3.2) the primary aberration of a hexapole element (a three-fold astigmatism) is compensated by a second hexapole arranged in an antisymmetric configuration coupled through a round lens transfer doublet. These two hexapoles generate a third order combination aberration with the correct symmetry to compensate the positive spherical aberration of the objective lens. In practice, additional weak multipole elements are also incorporated in the overall corrector design to correct residual aberrations arising from misalignment of the electron beam trajectory within the corrector. This design is able to correct for all third order aberrations over a field of view sufficient for CTEM imaging. Reviews detailing the development of aberration-correction can be found in Rose (2009) and Hawkes (2009).

The optical arrangement described above forms the basis of the CTEM corrector design produced commercially by CEOS GmbH (www.ceos-gmbh.de) and installed in many instruments worldwide with resolutions better than 0.1 nm at 200 kV (for example Sawada *et al.* (2005)). The important difference between uncorrected and aberration-corrected CTEM instruments are the additional corrector elements located immediately after the objective lens (Figure 9.1).

9.2.2 Related Instrumental Developments

In parallel with the development of aberration-correction, CTEM resolution limits have also been improved by reducing instabilities in the high voltage and objective lens supplies, by careful mechanical column design and consideration of environmental conditions within the microscope room. The resolution of aberration-corrected CTEM's is currently limited by chromatic aberration of the objective lens and by the energy spread in the electron beam. Monochromation improves the latter increasing the information limit and also provides enhanced spectroscopic performance. However, monochromation inevitably reduces the current in the electron beam. This implies that with a monochromator the practical optimum energy resolution for CTEM imaging is often limited by the requirement for sufficient electron dose to record images above the detector noise floor with reasonable exposure times such that specimen drift is not significant.

Direct correction of the chromatic aberration of the objective lens is therefore clearly desirable for CTEM imaging at high resolution and there are several ways in which suitable chromatic correctors could be constructed (Hawkes, 2007). The hexapole elements used in current CTEM correctors are not suitable for chromatic correction but there are optical geometries suitable for correction of both spherical and chromatic aberration in CTEM (Rose, 2009). A corrector based on a design by Rose has recently been successfully installed in a medium voltage microscope as part of the TEAM project (Haider *et al.*, 2009; Kisielowski *et al.*, 2008).

9.3 CTEM IMAGING THEORY

According to the theory of reciprocity discussed in Chapters 3, 5 and 6, bright field image formation in CTEM can be considered as the reciprocal of bright field image formation in STEM considering only elastic scattering (section 5.3.1). In this section we briefly summarise CTEM image formation in the context of aberration-correction. A complementary outline mathematical treatment is given in Box 9.1 but a full coverage of the theory of electron optics in CTEM is beyond the scope of this text and can be found elsewhere (Born and Wolf, 2002; Hawkes and Kasper, 1989; Reimer, 1997; Spence, 2002).

9.3.1 CTEM Image Formation

If it is assumed that the electron source is monochromatic, that the electron wave is perfectly coherent, and that all lenses are aberration free, rays scattered from a single point in an object will be recombined to a point in the image plane (Figure 9.2). Although not formally derived here it can be shown that the specimen, objective

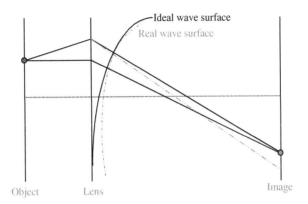

Figure 9.2   The effect of the aberrations of the objective lens modifying the wave surface in the back focal plane of the lens.

lens back focal and image planes are mathematically related by Fourier transform operations.

In a CTEM, the source produces an electron wave which passes through a series of condenser lenses leading to a plane wave incident on the specimen.[1] A thin sample can be assumed to act as a weak phase object (Box 9.1A). In this case the effect of the specimen is to produce a spatially varying phase shift in the electron wave such that at the specimen exit surface the phase of the electron wave contains direct information about the specimen potential.

Contrast in a high resolution CTEM image arises from the interference of the electron wave with itself. However, the recorded image contains only the intensity of the wave resulting from this interference and thus the phase information is lost. Suitable adjustment of the defocus produces an additional phase shift converting phase changes in the specimen to contrast in the image (details of optimal *phase-contrast imaging* conditions are given in section 9.4). Therefore an image of a weak phase object shows no contrast at zero defocus (the Gaussian defocus). The difficulties of phase contrast imaging in CTEM arise because this phase change introduced by the objective lens oscillates as a function of spatial frequency (section 9.3.2) A mathematical description of this imaging process in given in Box 9.1B.

---

[1] As previously noted in Chapters 1, 5 and 6, the final condenser lens in CTEM is the probe forming lens in the STEM imaging mode so in STEM it is termed the objective lens. Similarly, the CTEM condenser aperture is the limiting aperture in the STEM imaging mode so in STEM it is termed the objective aperture.

9.3.2 The Wave Aberration Function

For high resolution imaging the influence of the objective lens in CTEM can be conveniently formulated in terms of a wave aberration function, $\chi(\mathbf{K})$, (see Box 9.2) that describes the phase perturbation of the aberrated wave front compared to the ideal spherical wave (Figure 9.2 and section 4.2).

The overall effect of the wave aberration function is to increase the distance travelled by the electron wave which leads to a phase change. Therefore for phase contrast imaging in the CTEM the effect of the aberrations on the wavefunction, can be described by a phase factor $\exp[-i\chi(\mathbf{K})]$ as outlined in Box 9.1C.

Equation 9.2.2 in Box 9.2 is an approximation and in the general case the wave aberration function depends not only on the complex reciprocal space vector $\mathbf{K}$, but also on position in the image plane and on the energy of the electrons. Terms in $\chi(\mathbf{K})$ that have a position dependence are 'off-axial' aberrations and at high magnification (as used in high resolution CTEM imaging), where the field of view is small these can generally be neglected in the *isoplanatic* approximation[2]. The energy dependent terms in the wave aberration function are also frequently ignored in the *monochromatic* approximation since for field emission sources the energy variation within the electron beam is small (approximately 0.7 eV for a Schottky FEG and 0.3 eV for a cold FEG).

Box 9.1

9.1 a) Object Approximations

A *phase object* modulates the incident wave by a *specimen transmission function*, $\phi(\mathbf{R})$, where

$$\phi(\mathbf{R}) = \exp(+i\sigma V(\mathbf{R}))$$

and where σ is an interaction constant;

$$\sigma = 2\pi m e \lambda / h^2$$

and m is the relativistic mass. If the phase changes are small the transmission function can be further simplified for a *weak phase object* (WPO) as;

$$\phi(R) = \exp(+i\sigma V(\mathbf{R})) \sim 1 + i\sigma V(\mathbf{R})$$

---

[2] This position is likely to have to be reviewed in the future as larger detectors are introduced that increase the useable field of view.

This same approximation is used in STEM imaging and is discussed in sections 1.5 and 5.3.1.

9.1 b) Imaging Theory for a Perfect Lens

For an incident electron beam represented by the plane wave, ψ_o, the amplitude at the exit face of the crystal, $\psi_{exit}(\mathbf{R})$, is;

$$\psi_{exit}(\mathbf{R}) = \psi_o\phi(\mathbf{R})$$

where $\phi(\mathbf{R})$ is the complex *specimen transmission function* used in section 9.1A. For Fraunhofer diffraction the amplitude in the back-focal plane (the diffraction plane) is obtained by taking the Fourier transform of $\psi_{exit}(\mathbf{R})$ to give;

$$\tilde{\psi}_{exit}(\mathbf{K}) = \mathbf{F}(\psi_o\phi(\mathbf{R}))$$

Where a Fourier transform is denoted by the symbol $\mathbf{F}$. The wavefunction in the image plane, $\psi_{image}(\mathbf{R})$, is then given by a further Fourier transform;

$$\psi_{image}(\mathbf{R}) = \mathbf{F}\{\mathbf{F}(\psi_o\phi(\mathbf{R}))\} = \psi_o\phi(R)$$

The recorded image intensity is then given by the product of the image wavefunction and its complex conjugate;

$$I_{image}(\mathbf{R}) = \psi_{image}(\mathbf{R})\psi_{image}{}^*(\mathbf{R}) = |\psi_o|^2$$

9.1 c) Imaging Theory including Objective Lens Aberrations

The effects of the objective lens aberrations (section 9.3.2), partial coherence and the finite size of the objective aperture (section 9.3.3) can be included in the above mathematical treatment using a transfer function[3], $A(\mathbf{K})$ where;

$$A(\mathbf{K}) = H(\mathbf{K})\exp[-i\chi(\mathbf{K})]$$

As already described, the wave aberration function, $\chi(\mathbf{K})$, describes the effect of the objective lens aberrations (Box 9.2 and section 4.2) and $H(\mathbf{K})$ describes the effects of the objective aperture and the partial coherence of the electron beam unlike the notation in Chapters 4 and 5 where $H(\mathbf{K})$ is simply a probe limiting aperture function. Where the objective aperture is sufficiently large that it can be ignored, $H(\mathbf{K})$ can be approximated for a weak object as the

---

[3]Here we will use $A(\mathbf{K})$ to represent the transfer function of the lens for consistency with other chapters in this book. Note in the case of HRTEM imaging, $A(\mathbf{K})$ includes the effects of the lens aberrations, the objective aperture aberration as well as partial coherence, the latter being described in the form of a partial coherence envelope (see section 9.3.3 and Box 9.3).

product of envelope functions describing the partial spatial and temporal coherence (see box 9.3). Thus the overall effect of A(K) is to modify the amplitude distribution in the back-focal plane of the objective lens as $\tilde{\psi}_{exit}$ (K)A(K). Hence the amplitude distribution in the image plane is given by;

$$\psi_{image}(\mathbf{R}) = \mathbf{F}\{\tilde{\psi}_{exit}(\mathbf{K})A(\mathbf{K})\} = \psi_0 \exp(+i\sigma V(\mathbf{R})) \otimes \tilde{A}(\mathbf{R})$$

where $\tilde{A}(\mathbf{R})$ is the Fourier transform of A(K) and $\otimes$ denotes a convolution operation. Using $\exp(-i\chi(\mathbf{K})) = \cos\chi(\mathbf{K}) - i\sin\chi(\mathbf{K})$, the amplitude distribution in the image plane for a *weak phase object* is given by;

$$\psi_{image}(\mathbf{R}) = \psi_0[1 + i\sigma V(\mathbf{R}) \otimes \mathbf{F}\{H(\mathbf{K})\cos\chi(\mathbf{K})\} + \sigma V(\mathbf{R}) \otimes \mathbf{F}\{H(\mathbf{K})\sin\chi(\mathbf{K})\}]$$

Giving an image intensity;

$$I(\mathbf{R}) = \psi_{image}(\mathbf{R})\psi_{image}^*(\mathbf{R}) = 1 + 2\sigma V(\mathbf{R}) \otimes \mathbf{F}\{H(\mathbf{K})\sin\chi(\mathbf{K})\}$$

assuming that terms of order $\sigma^2 V^2$ can be neglected. From this result it is clear that for these approximations only the imaginary part of A(K) contributes to the intensity and is therefore known as the *Phase Contrast Transfer Function*, PCTF(K),

$$PCTF(\mathbf{K}) = H(\mathbf{K})\sin\chi(\mathbf{K})$$

The shape of this function for a typical corrected and uncorrected values of spherical aberration is shown in Figure 9.3. See a corresponding description for STEM bright field imaging in section 4.2.1.

Box 9.2

The Wave Aberration Function

For an uncorrected CTEM the wave aberration function, $\chi(\mathbf{K})$, is dominated by the effects of spherical aberration ($C_{3,0}$) and defocus ($C_{1,0}$) so that for a well aligned instrument other aberrations can be ignored. This gives the wave aberration function the relatively simple form;

$$\chi(\mathbf{K}) = \left\{\frac{2\pi}{\lambda}\right\}\left\{\frac{1}{2}C_{1,0}\lambda^2\mathbf{K}^*\mathbf{K} + \frac{1}{4}C_{3,0}\lambda^4\mathbf{K}^{*2}\mathbf{K}^2\right\} \qquad (9.2.1)$$

As discussed in Chapter 4, section 4.2, the seemingly counter-intuitive notation (e.g. $C_{1,0}$ for a second order term in $\chi(\mathbf{K})$) stems from the ray-optical theory of Seidel aberrations, which are described in terms of displacements of ray-path intersections with the image plane. These displacements are proportional to the gradient of the wave aberration function, hence an Nth order Seidel aberration will correspond to a term of order $N + 1$ in the wave aberration function. The aberration due to defocus, $C_{1,0}$, is linear in angle with respect to the image aberration function and is hence described as a first order aberration, whereas the aberration due to spherical aberration $C_{3,0}$ is cubic in the image aberration function and is hence described as a third order aberration.

For aberration-corrected instruments the wave aberration function equation is more complicated because there are many aberration coefficients whose contributions to the phase shift have approximately equal weight. Taylor expansion to sixth order in K about the origin of zero scattering angle gives the wave aberration function, in terms of all important aberration coefficients for corrected instruments up to sixth order in K (as listed in Table I of Appendix 1) as;

$$
\begin{aligned}
\chi(\mathbf{K}) = \mathrm{Re} \left\{ \frac{2\pi}{\lambda} \right. & \left\{ C_{0,1}\lambda\mathbf{K}^* + \frac{1}{2}C_{1,2}\lambda^2\mathbf{K}^{*2} + \frac{1}{2}C_{1,0}\lambda^2\mathbf{K}^*\mathbf{K} \right. \\
& + \frac{1}{3}C_{2,3}\lambda^3\mathbf{K}^{*3} + \frac{1}{3}C_{2,1}\lambda^3\mathbf{K}^{*2}\mathbf{K} \\
& + \frac{1}{4}C_{3,4}\lambda^4\mathbf{K}^{*4} + \frac{1}{4}C_{3,2}\lambda^4\mathbf{K}^{*3}\mathbf{K} + \frac{1}{4}C_{3,0}\lambda^4\mathbf{K}^{*2}\mathbf{K}^2 \\
& + \frac{1}{5}C_{4,5}\lambda^5\mathbf{K}^{*5} + \frac{1}{5}C_{4,3}\lambda^5\mathbf{K}^{*4}\mathbf{K} + \frac{1}{5}C_{4,1}\lambda^5\mathbf{K}^{*3}\mathbf{K}^2 \\
& + \frac{1}{6}C_{5,6}\lambda^6\mathbf{K}^{*6} + \frac{1}{6}C_{5,2}\lambda^6\mathbf{K}^{*4}\mathbf{K}^2 \\
& \left. + \frac{1}{6}C_{5,0}\lambda^6\mathbf{K}^{*3}\mathbf{K}^3 + +\frac{1}{6}C_{5,4}\lambda^6\mathbf{K}^{*5}\mathbf{K} \right\} \left. \right\}
\end{aligned}
\tag{9.2.2}
$$

The first, second and third terms in equation (9.2.2) above are readily adjustable in all TEMs and describe position in the image plane, two fold astigmatism and defocus respectively. The wave aberration function can be presented in several different forms. Here we have chosen to give it with reference to the spatial frequency vector, $\mathbf{K}$, but it can equally be presented with reference to a complex scattering angle $w = \mathbf{K}\lambda$;

$$
\begin{aligned}
\chi(\mathbf{K}) = \mathrm{Re} \left\{ \frac{2\pi}{\lambda} \right. & \left\{ C_{0,1}w^* + \frac{1}{2}C_{1,2}w^{*2} + \frac{1}{2}C_{1,0}w^*w \right. \\
& + \frac{1}{3}C_{2,3}w^{*3} + \frac{1}{3}C_{2,1}w^{*2}w
\end{aligned}
$$

$$+ \frac{1}{4}C_{3,4}w^{*4} + \frac{1}{4}C_{3,2}w^{*3}w + \frac{1}{4}C_{3,0}w^{*2}w^2$$

$$+ \frac{1}{5}C_{4,5}w^{*5} + \frac{1}{5}C_{4,3}w^{*4}w + \frac{1}{5}C_{4,1}w^{*3}w^2$$

$$+ \frac{1}{6}C_{5,6}w^{*6} + \frac{1}{6}C_{5,2}w^{*4}w^2 + \frac{1}{6}C_{5,0}w^{*3}w^3 + + \frac{1}{6}C_{5,4}w^{*5}w \Big\}\Big\}$$

An alternative presentation which makes clear the angular dependence was given in Chapter 4 equation (4.2). More detail on the aberration coefficients and the definition of the wave aberration function is given in Appendix 1.

Box 9.3

The Partial Coherence Envelopes

Assuming the electron source has a Gaussian intensity profile the partial spatial coherence envelope can be written as;

$$E_s(\mathbf{K}) = \exp\left[\left(\frac{-\alpha^2}{4\lambda^2}\right)\left(\frac{\partial \chi(\mathbf{K})}{\partial \mathbf{K}}\right)^2\right]$$

where α is the semi-angle characterising the Gaussian distribution (beam convergence). Similarly the partial temporal coherence envelope can be written as;

$$E_c(\mathbf{K}) = \exp[-\tfrac{1}{2}(\pi\lambda\Delta)^2\mathbf{K}^4)]$$

Where Δ is the standard deviation of the Gaussian focal spread distribution resulting from the instabilities in the energy of the electron source (ΔE), the accelerating voltage (ΔV), and in the objective lens power supply (ΔI);

$$\Delta = C_c\sqrt{\left(\frac{\Delta V}{V}\right) + \left(\frac{\Delta E}{E}\right) + 4\left(\frac{\Delta I}{I}\right)}$$

Overall the *Phase Contrast Transfer Function*, PCTF(K) including the effect of partial coherence for CTEM imaging is then given by;

$$\text{PCTF}(\mathbf{K}) = E_s(\mathbf{K})E_c(\mathbf{K})\sin\chi(\mathbf{K})$$

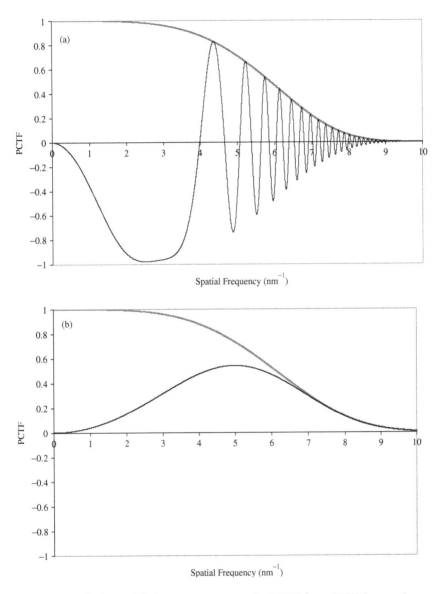

Figure 9.3   The lower black curve represents the PCTF for a CTEM operating at 200 kV including the effect of partial coherence for (a) a typical uncorrected value of spherical aberration, $(C_{3,0} = 1$ mm $C_{1,0} = -50$ nm) and (b) for a typical aberration-corrected value of spherical aberration $(C_{3,0} = -10\,\mu$m $C_{1,0} = 5$ nm). The upper grey curve represents the combined partial and spatial coherence envelope. Note the negative sign of the corrected spherical aberration and the corresponding over focus condition.

9.3.3 Partial Coherence

For perfect coherence, CTEM resolution will be limited by the semi angle of the objective aperture in the back focal plane. This can be described mathematically using a simple circular function which has a value of 1 within the objective aperture and 0 outside. For high resolution CTEM it is reasonable to assume that that the objective aperture is sufficiently large (or absent) that its effect can be ignored. Thus practical CTEM resolution is limited by effects due to partial coherence.

The coherence of the electron beam determines the extent to which electron waves can interfere so as to produce a high resolution CTEM image. Thus, prior to considering optimum imaging conditions for aberration-corrected CTEM it is necessary to understand the effects of partial coherence on the recorded image contrast. This requires a different approach to that used for partial coherence in ADF STEM imaging (Chapter 5) since phase contrast CTEM imaging is a coherent imaging mode. A comprehensive description of partial coherence can be found in Born and Wolf (2002), Reimer (1997) and Kirkland *et al.* (2008).

There are two components contributing to partial coherence: partial *spatial* coherence, and partial *temporal* coherence introduced in Chapter 1, section 1.2. The combined effect of these is to restrict the information in high-resolution CTEM images.

The finite size of the electron source results in a loss of *spatial* coherence since each point on the specimen is effectively illuminated from a range of different directions. This effect can be treated by assuming a Gaussian intensity profile about the average incident beam direction. Using this approximation the beam convergence is given by the root-mean-squared spread of the Gaussian beam profile with typical values of 0.1 mrad for a field emission source.

Partial *temporal* coherence originates from the combined effects of a finite energy distribution in the electron source and fluctuations in both the accelerating voltage and the objective lens current. This can be considered as a smearing of the nominal defocus, which can also be approximated as a Gaussian distribution with a standard deviation referred to as the focal spread (Δ). This focal spread can be calculated from the fluctuations in the accelerating voltage, the objective lens current, and the energy of the electrons leaving the source, together with the magnitude of the chromatic aberration coefficient of the objective lens and is typically several nanometres in modern CTEM instruments (Box 9.3).

The detector system also introduces an spatial frequency-dependent attenuation of the image contrast which can be included in the imaging process through a Modulation Transfer Function, (MTF(K)) (Meyer and Kirkland, 2000). This function has an effect similar to the partial coherence envelopes, except that the limiting spatial frequency depends on magnification. Hence unlike partial coherence the detector transfer at a particular spatial frequency is improved at a high magnification albeit at the expense of a reduced field of view.

9.4 CORRECTED IMAGING CONDITIONS

For phase contrast CTEM imaging the recorded image intensities are strongly affected by phase shifts arising from the objective lens aberrations as described in section 9.3.2. Consequently, in contrast to HAADF STEM images, high resolution CTEM images often show contrast reversals and delocalisation and can therefore be difficult to interpret except under specific imaging conditions. Compared to uncorrected CTEM imaging, image delocalisation is greatly reduced for aberration-corrected imaging conditions as illustrated in Figure 9.4. However,

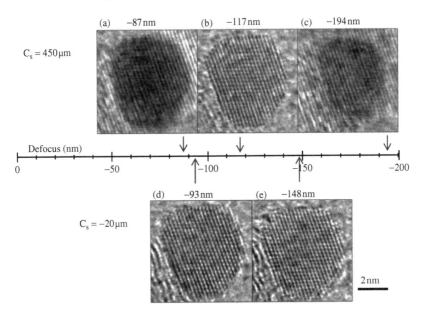

Figure 9.4   CTEM images of a gold particle on a carbon film acquired using a conventional uncorrected value for the spherical aberration (a,b,c) and using an aberration-corrected condition (d, e). The defocus value is given above each image and also referred to on the linear scale.

Table 9.1 Aberration-corrected imaging conditions modified from Chang, Kirkland and Titchmarsh (2006). Values of the defocus ($C_{1,0}$) and the spherical aberration coefficient ($C_{3,0}$) are given for a CTEM operating at 200 kV with a RMS focal spread distribution (Δ) of 4 nm and with a fixed fifth order spherical aberration ($C_{5,0}$) value of 2 mm except where otherwise stated.

Imaging condition	Aberration coefficients	
	Definition	Magnitude
Phase contrast Scherzer defocus condition	$C_{1,0} = -(C_{3,0}\lambda)^{1/2}$	if $C_{3,0} = 1$ mm $C_{1,0} = -50$ nm
Phase contrast $C_{5,0}$ limited imaging	$C_{3,0} = \mp 2.88(C_{5,0}^2\lambda)^{1/3}$ $C_{1,0} = \pm 1.56(C_{5,0}\lambda^2)^{1/3}$	$C_{3,0} = \mp 6\,\mu$m $C_{1,0} = \pm 4$ nm
Phase contrast C_c limited imaging	$C_{3,0} = \pm (\pi\Delta)^2/\lambda$ $C_{1,0} = \mp \pi\Delta$	$C_{3,0} = \pm 63\,\mu$m $C_{1,0} = \mp 13$ nm
Amplitude contrast $C_{5,0} = 0$	$C_{3,0} = 0$ $C_{1,0} = 0$	$C_{3,0} = 0\,\mu$m $C_{1,0} = 0$ nm

delocalisation and contrast reversals are not completely removed by aberration-correction and an understanding of the imaging conditions is still necessary in order for high resolution images to be interpretable.

For uncorrected CTEM, there are well known imaging conditions where the objective lens defocus is adjusted to partially compensate the phase shifts due to third order spherical aberration (as discussed in Chapter 3 and section 4.6). The most commonly used of these is the Scherzer defocus[4], which maximises the range of spatial frequencies for which the PCTF is close to -1 ($\chi(\mathbf{K}) \approx (-\pi/2)$) (as illustrated in Figure 9.3).

However, in an aberration-corrected microscope where spherical aberration is adjustable and phase contrast imaging conditions must take this into account. This leads to several different optimal settings in which defocus and spherical aberration are adjusted to compensate for the effect of either 5th order spherical aberration or chromatic aberration (Table 9.1).

9.4.1 The Use of Negative Spherical Aberration

For an aberration-corrected CTEM the third order spherical aberration is adjustable within a range that includes negative values (typically

---

[4] As an alternative the 'extended Scherzer defocus', $C_{1,0} = -1.2(C_{3,0}\lambda)^{1/2}$ achieves a slightly better resolvable distance by allowing the PCTF passband to contain a local minima.

$\pm 20\,\mu m$ at $200\,kV$). If the sign of spherical aberration is reversed the value for defocus must also be reversed so that an overfocus condition is used when the spherical aberration is negative. This gives rise to a PCTF that is mirrored with respect to the traditional positive spherical aberration and underfocus condition (Figure 9.3). For linear imaging, where interferences between diffracted beams are neglected, these two conditions are symmetric and give equal but reversed contrast.

However, for moderately thick specimens (typically less than $10\,nm$) negative spherical aberration imaging (NCSI) gives improved contrast at high resolution. It has been shown that this effect is due to non-linear interferences adding to the linear image contrast for the negative spherical aberration condition but reducing the linear contrast when the spherical aberration is positive (Jia and Urban, 2004). NCSI has been used successfully to increase the contrast of atomic columns of light elements such as nitrogen in Si_3N_4 or GaN and oxygen in ceramic oxides as shown in Figure 9.5(a).

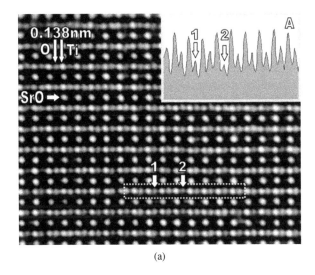

(a)

Figure 9.5   (a) Experimental aberration-corrected CTEM image of $SrTiO_3$ [011] acquired using the negative spherical aberration imaging condition ($C_{3,0} = -40\,\mu m$, $C_{1,0} = 8\,nm$). Atom columns in the $4\,nm$ thick specimen appear bright on a dark background, and all atomic positions including pure oxygen columns are resolved. The oxygen-atom contrast is weaker at positions 1 and 2 as shown quantitatively by the intensity trace in inset (A). This trace has been found to accurately match simulations where the percentage oxygen concentration is reduced to 85 and 80%, respectively. (Reproduced from Jia *et al.*, *Science*, 299, 870–871 (2003).)

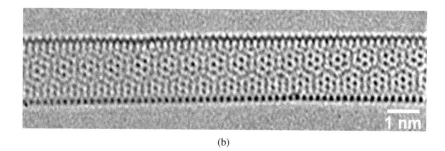

(b)

Figure 9.5   (b) Experimental-aberration corrected CTEM image acquired at 80 kV showing a single walled carbon nanotube. Image courtesy of Jamie Warner (University of Oxford).

9.4.2 Amplitude Contrast Imaging

The most intuitive approach to aberration-corrected imaging is to set all aberrations to zero (as is done for corrected HAADF STEM) giving rise to pure amplitude contrast which in linear imaging theory is proportional to $\cos(\chi(\mathbf{K}))$ – the amplitude contrast transfer function. However, amplitude contrast is weak for thin specimens and images are highly sensitive to the accuracy of the aberration compensation, so in practice images often contain a mixture of phase and amplitude contrast. As a result this imaging condition has not been widely adopted and most optimal imaging conditions maximise phase contrast.

9.5 ABERRATION MEASUREMENT

As discussed in section 4.5 for STEM imaging, practical aberration-correction depends on the accuracy of aberration measurement. For CTEM imaging, aberration measurement relies on the fact that tilting the incident illumination shifts the origin of the wave aberration function and thus tilt induced changes in the experimentally measurable aberrations can be related to the axial aberration coefficients (Haigh et al., 2009).

9.5.1 Aberration Measurement From Image Shifts

The displacement between two images acquired at different beam tilts depends on all the aberration coefficients, which can thus be determined by minimising the least square difference between measured and expected image shifts for a set of known beam tilts. Experimental image shifts can be measured, from the position of the peak in the cross-correlation function calculated between two images. However, for high resolution imaging, it is necessary to compensate for peak distortions introduced by higher order aberrations and this requires *a priori* knowledge of the approximate axial aberrations. Image shift measurements also fail for periodic specimens, where positions differing by an integer number of lattice vectors cannot be distinguished. Most seriously, the tilt-induced displacement is indistinguishable from specimen drift and therefore this approach has been limited to microscope alignment at low resolution.

9.5.2 Aberration Measurement from Diffractograms

A more commonly used approach relies on the determination of tilt induced changes in defocus, and/or two fold astigmatism. Measurements are made using diffractograms (power spectra) of the recorded experimental images acquired for a number of known illumination tilts (Figure 9.6). Although this approach (known as a Zemlin tableau) necessitates the presence of some amorphous material, (either as a result of specimen damage, contamination or the presence of a carbon support film) these measurements have been successfully automated by matching experimental data to simulations. For high resolution imaging the

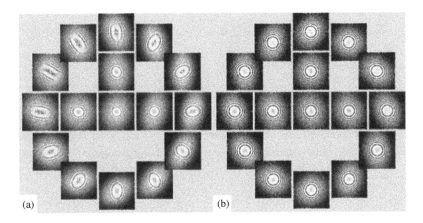

(a)   (b)

Figure 9.6 Zemlin tableaux of diffractograms (a) before and (b) after aberration-correction. (Reproduced from John L. Hutchison, John M. Titchmarsh, David J.H. Cockayne, Ron C. Doole, Crispin J.D. Hetherington, Angus I. Kirkland, H. Sawada. A versatile double aberration-corrected, energy filtered HREM/STEM for materials science, *Ultramicroscopy*, 103 7–15© (2005) with permission from Elsevier.)

major benefits of this method compared to image shift measurement are improved accuracy and insensitivity to specimen drift.

9.5.3 An Alternative Approach to Aberration Measurement

An alternative approach to aberration measurement has also been developed which is less reliant on the presence of amorphous material whilst retaining sufficient accuracy for high resolution application (Meyer et al., 2002). This approach uses a phase compensated correlation function, PCF_W, which is related to the conventional cross correlation function. However, unlike the cross correlation function, which is periodic in the presence of strong crystal reflections, the PCF_W consists of a single peak. The shape of this peak is related to the focus difference between two images.

From the values of relative defoci determined using the PCF_W an initial estimate of the wavefunction *in the plane of a reference image* can be made and the *absolute* values of defocus and two-fold astigmatism can subsequently be determined using a *Phase Contrast Index Function*, f_{PCI} (Meyer et al., 2002). Therefore using short focal series acquired at a number of different illumination tilts, this multistep method can determine the tilt induced changes in defocus and two-fold astigmatism and thus all the aberration coefficients (Meyer et al., 2004).

9.6 INDIRECT ABERRATION COMPENSATION

Post acquisition image processing allows for indirect aberration compensation in CTEM in an approach known as exit wave restoration. This technique uses a series of experimental images and measured aberrations to recover the full complex electron wave which for a thin specimen is directly interpretable in terms of the atomic structure of the specimen. A typical data set consists of a short focal series of 10–20 images and this approach is now implemented in a number of commercial software packages.

Combining direct aberration-correction and exit wave restoration has a number of advantages, including compensation of higher order aberrations which cannot be corrected electron optically (Kirkland et al., 2006). This technique has been successfully applied to the imaging of catalytic metal nanoparticles (Cervera *et al.*, 2007) and defects in semiconductor materials (Tillmann *et al.*, 2004).

9.7 ADVANTAGES OF ABERRATION-CORRECTION FOR CTEM

For CTEM imaging the primary benefit of aberration-correction is improved interpretable resolution to much better than 0.1 nm for medium voltage instruments. However, there is increasingly an interest in high resolution imaging at lower voltages (80 kV and below) for studies of radiation sensitive materials. This is particularly important for carbon based materials such as fullerenes, nanotubes and graphene where CTEM imaging at close to 0.1 nm resolution as been achieved (Figure 9.5(b)).

As well as improving the interpretable resolution, spherical aberration-correction also reduces the tilt induced changes in the lower order aberration coefficients as demonstrated in tilt tableaux of diffractograms in Figure 9.6. Urban *et al.* (2009) have exploited this effect using beam tilt for precise zone axis alignment. This effect can also be exploited to relax the requirement for parallel illumination in CTEM imaging allowing a small amount of beam convergence to be used thus increasing the beam current. This is particularly important for energy filtered CTEM imaging where the signal to noise ratio is critical, and in this way aberration-corrected energy filtered CTEM imaging using the low loss region of the spectrum has been used for elemental mapping of nanometre size precipitates (Lozano-Perez *et al.*, 2009).

9.8 CONCLUSIONS

Aberration-correction for CTEM is now fully developed and directly interpretable atomic resolution images can be routinely obtained often providing information that is complementary to aberration-corrected STEM imaging. Turning spherical aberration into a variable has required a re-evaluation of established definitions of optimal imaging conditions. It has been shown that over-correction of the spherical aberration to a small negative value gives enhanced image contrast and this has been applied successfully to materials with light and heavy elements in close proximity. Exit wave restoration can also be performed using aberration-corrected images allowing retrieval of the complex electron wave, further improving structural interpretation.

Acknowledgements

We acknowledge financial support from the EPSRC (grants EP/F048009/1, GR/S83968/01) and from the European Union for the Framework 6 programme under a contract for an Integrated Infrastructure Initiative (Reference 026019 ESTEEM).

REFERENCES

Born, M. and Wolf, E. (2002) *Principles of Optics*, 7th edn, Cambridge University Press.

Cervera, L., Chang, L.Y., Kirkland, A.I., Hetherington, C., Ozkaya, D. and Dunin-Borkowski, R.E. (2007) Aberration-Corrected Imaging of Active Sites On Industrial Catalyst Nanoparticles, *Angewandte Chemie Int. Ed.*, 46, 3683–3685.

Chang, L.Y., Kirkland, A.I. and Titchmarsh, J.M. (2006) On the Importance of Fifth-Order Spherical Aberration for a Fully Corrected Electron Microscope, *Ultramicroscopy*, 106, 301–306.

Haider, M., Rose, H., Uhlemann, S., Schwan, E., Kabius, B. and Urban, K. (1998) A Spherical-Aberration-Corrected 200kV Transmission Electron Microscope, *Ultramicroscopy*, 75, 53–60.

Haider, M., Hartel, P., Müller, H., Uhlemann, S. and Zach, J. (2009) Current and Future Aberration Correctors for the Improvement of Resolution in Electron Microscopy, *Phil. Trans. R. Soc. A.*, September 28, 2009 367, 3665–3682; doi:10.1098/rsta.2009.0121.

Hawkes, P.W. and Kasper, E. (1989, 1994, 1996) *Principles of Electron Optics*, Academic Press, London.

Hawkes, P.W. (2007) Aberration Correction, in: P.W. Hawkes and J. C. H. Spence (eds), *Science of Microscopy*, Vol. 1, Springer, New York, p. 696–748.

Hawkes, P.W. (2009) Aberration Correction Past and Present *Phil. Trans. R. Soc. A.* September 28, 2009 367: 3637–3664; doi:10.1098/rsta.2009.0004.

Haigh, S. J., Sawada, H., Kirkland, Angus I. (2009) Optimal Tilt Magnitude Determination for Aberration-Corrected Super Resolution Exit Wave Function Reconstruction, *Phil. Trans. R. Soc. A.*, September 28, 2009 367, 3755–3771.

Jia, C.L. and Urban, K. (2004) Atomic-Resolution Measurement of Oxygen Concentration in Oxide Materials, *Science*, 303, 5666, 2001–2004.

Kirkland, A.I., Chang, L.Y. and Hutchison, J.L. (2007) Atomic Resolution Transmission Electron Microscopy, in: P. Hawkes, J. C. H. Spence (eds), *Science of Microscopy*, Vol. I, Springer, New York, p. 3–64.

Kirkland, A.I., Meyer, R.R. and Chang, L.Y. (2006) Local Measurement and Computational Refinement of Aberrations for HRTEM, *Microscopy And Microanalysis*, 12, 6, 461–468.

Kirkland, A.I., Nellist, P.D., Chang, L.-Y. and Haigh, S.J. (2008) Aberration-Corrected Imaging in Conventional Transmission Electron Microscopy and Scanning Transmission Electron Microscopy, *Advances in Imaging and Electron Physics*, 153, Elsevier, Ch 8 pp. 253–325.

Kisielowski, C. *et. al.* (2008) Detection of Single Atoms and Buried Defects in Three Dimensions by Aberration-Corrected Electron Microscope With 0.5-angstrom Information Limit, *Microscopy and Microanalysis*, 14, 469–477.

Lozano-Perez, S., de Castro, V. and Nicholls, R. (2009) Achieving Sub-Nanometre Particle Mapping With Energy-Filtered TEM, *Ultramicroscopy*, 109(10), 1217–1228.

Meyer, R.R. and Kirkland, A.I. (2000) Characterisation of the Signal and Noise Transfer of CCD Cameras for Electron Detection, *Microsc. Res. Techniq.*, 49, 269–280.

Meyer, R.R, Kirkland, A.I. and Saxton, W.O (2002) A New Method for the Determination of the Wave Aberration Function for High-Resolution TEM. 1. Measurement of the Symmetric Aberrations, *Ultramicroscopy*, 92, 89–109.

Meyer, R., Kirkland, A. and Saxton, W. (2004) A New Method for the Determination of the Wave Aberration Function for High-Resolution TEM. 2. Measurement of the Antisymmetric Aberrations, *Ultramicroscopy*, 99, 115–123.

Reimer, L. (1997) *Transmission Electron Microscopy: Physics of Image Formation and Microanalysis*, 4th edn, Springer Series in Optical Sciences, Springer, Berlin; New York.

Rose, H.H. (2009) Historical aspects of aberration correction *Journal of Electron Microscopy*, 58(3): 77–85 doi: 10.1093/jmicro/dfp012.

Sawada, H., Tomita, T., Naruse, M., Honda, T., Hambridge, P., Hartel, P., Haider, M., Hetherington, C., Doole, R., Kirkland, A., Hutchison, J., Titchmarsh, J. and Cockayne, D. (2005) Experimental Evaluation of a Spherical Aberration-Corrected TEM and STEM, *Journal of Electron Microscopy*, 54, 119–121.

Smith, D.J. (1997) The Realization of Atomic Resolution with the Electron Microscope, *Reports on Progress in Physics* 60 1513–1580.

Spence, J. (2002) *High Resolution Electron Microscopy*, 3rd edn, Oxford University Press, Oxford.

Tillmann, K., Thust, A. and Urban, K. (2004) Spherical Aberration Correction in Tandem With Exit-Plane Wave Function Reconstruction: Interlocking Tools for the Atomic Scale Imaging of Lattice Defects in GaAs, *Microscopy And Microanalysis*, 10(2), 185–198.

Urban, K.W., Jia, C.L., Houben, L., Lentzen, M., Mi, S.-B., and Tillmann, K. (2009) Negative Spherical Aberration Ultrahigh-Resolution Imaging in Corrected Transmission Electron Microscopy, *Phil. Trans. R. Soc. A.*, September 28, 2009 367: 3735–3753; doi:10.1098/rsta.2009.0134.

Appendix A

Aberration Notation

There is no generally accepted notation for the aberration coefficients. In this book we have used the notation of Krivanek *et al.* (1999) who distinguish the aberration coefficients by means of two subscripts, the first denoting the order of the aberration (in terms of the wavevector $\mathbf{K}$) and the second, the (azimuthal) symmetry about the optic axis (as indicated in Table A.1 below). A further subscript label (a, b) is needed to separate orthogonal contributions to the same aberration when these are present.

Considering a typical round electron lens operating at 100 kV with a 5 mm gap and a focal length of 2 mm, the third order spherical aberration coefficient ($C_{3,0}$) will have an uncorrected value of a few millimetres, the fifth order spherical aberration coefficient ($C_{5,0}$) a value of several centimetres and the seventh order spherical aberration coefficient ($C_{7,0}$) a value of up to a metre. The uncorrected chromatic aberration coefficient C_c will be of a similar order of magnitude to $C_{3,0}$.

This notation has been widely adopted by the STEM community but differs from the Saxton notation (Typke and Dierksen (1997)) which is commonly used in CTEM. Aberration correctors produced by CEOS use the Uhlemann & Haider (1998) notation which looks the same as the Saxton notation but differs in some of its prefactors. Table A.2

Aberration-Corrected Analytical Transmission Electron Microscopy, First Edition.
Edited by Rik Brydson.
© 2011 John Wiley & Sons, Ltd. Published 2011 by John Wiley & Sons, Ltd.

Table A.1 The name, order in **K** and azimuthal symmetry for the aberration coefficients.

Aberration Coefficient	Krivanek notation	Order in K	Azimuthal symmetry
Image shift	$C_{0,1}$	1	1
Two fold astigmatism	$C_{1,2}$	2	2
Defocus (over focus positive)	$C_{1,0}$	2	inf
Three fold astigmatism	$C_{2,3}$	3	3
Axial coma	$C_{2,1}$	3	1
Four fold astigmatism	$C_{3,4}$	4	4
Axial star aberration	$C_{3,2}$	4	2
Spherical aberration	$C_{3,0}$	4	inf
Five fold astigmatism	$C_{4,5}$	5	5
Fourth order axial coma	$C_{4,1}$	5	1
Three lobe aberration	$C_{4,3}$	5	3
Six fold astigmatism	$C_{5,6}$	6	6
Fifth order rosette aberration	$C_{5,4}$	6	4
Fifth order axial star aberration	$C_{5,2}$	6	2
Fifth order spherical aberration	$C_{5,0}$	6	inf
Seven fold astigmatism	$C_{6,7}$	7	7
Sixth order axial coma	$C_{6,1}$	7	1
Sixth order three lobe aberration	$C_{6,3}$	7	3
Sixth order pentacle aberration	$C_{6,5}$	7	5
Eight fold astigmatism	$C_{7,8}$	8	8
Seventh order hexagon aberration	$C_{7,6}$	8	6
Seventh order rosette aberration	$C_{7,4}$	8	4
Seventh order star aberration	$C_{7,2}$	8	2
Seventh order spherical aberration	$C_{7,0}$	8	inf

shows the relation between the notations used by Uhlemann and Haider (1998), Saxton (1995), Krivanek *et al.* (1999) and Hawkes & Kasper (1989).

Practical comparison of the different notations is further complicated due to the variations that exist between the orientation of axes and the definition of the aberration function. In this book we have chosen to define χ as a phase shift due to aberrations but Uhlemann and Haider define χ as a distance resulting in a $2\pi/\lambda$ discrepancy.

Table A.2 The different notations for the aberration coefficients (this table includes terms not present in the publications cited, supplied by the authors in question – reproduced from Hawkes, 2008).

Uhlemann and Haider (1998)	Saxton (1995)	Krivanek *et al.* (1999)	Hawkes and Kasper (1989)
C_1	C_1	C_1	
A_1	A_1	$C_{1,2}$	$b_1 + i b_2$
B_2	$\frac{1}{3}\bar{B}_2$	$\frac{1}{3}C_{2,1}^*$	$\frac{1}{4}(3A_{30} - 3i A_{03} + A_{12} - i A_{21})$
A_2	A_2	$C_{2,3}$	$\frac{3}{4}(3A_{30} - 3i A_{03} + A_{12} - i A_{21})$
$C_s = C_3$	C_3	$C_{3,0} = C_3$	
S_3	$\frac{1}{4}\bar{B}_3$	$\frac{1}{4}C_{3,2}^*$	
A_3	A_3	$C_{3,4}$	
B_4		$\frac{1}{5}C_{4,1}^*$	
D_4		$\frac{1}{5}C_{4,3}^*$	
A_4		$C_{4,5}$	
C_5		$C_{5,0} = C_5$	
S_5		$\frac{1}{6}C_{5,2}^*$	
A_5		$C_{5,6}$	
R_5		$\frac{1}{6}C_{5,4}^*$	

REFERENCES

Hawkes, P.W. (2008) Aberrations in: J. Orloff (ed.), *Handbook of Charged Particle Optics*. CRC Press, Baton Rouge & London.

Hawkes, P.W. and Kasper, E. (1989) *Principles of Electron Optics*, Academic Press, London.

Kirkland, A.I., Nellist, P.D., Chang, L.-Y. and Haigh, S.J. (2008) Aberration-Corrected Imaging in Conventional Transmission Electron Microscopy and Scanning Transmission Electron Microscopy, in: *Advances in Imaging and Electron Physics*, 153, Ch 8 pp. 253–325 Elsevier.

Krivanek, O.L., Dellby, N. and Lupini, A.R. (1999) Towards Sub-Å Electron Beams, *Ultramicroscopy*, 78, 1–4, 1–11.

Typke, D. and Dierksen, K. (1995) Determination of Image Aberrations in High-Resolution Electron-Microscopy using Diffractogram and Cross-Correlation Methods, *Optik*, 99, 4, 155–166.

Uhlemann, S. and Haider, M. (1998) Residual Wave Aberrations in the First Spherical Aberration Corrected Transmission Electron Microscope, *Ultramicroscopy*, 72, 109.

Appendix B

General Notation

SYMBOLS

Note: **bold** notation refers to a vector and a capitalised vector refers to the transverse component of a three-dimensional vector (e.g., $\mathbf{K}$ and $\mathbf{k}$ for the case an electron wavevector).

$*$	a superscript indicates the complex conjugate of a function
$\| \|$	denotes the modulus of a function (also sometimes written as a non-bold vector)
$\angle$	denotes the phase of function
a	atom spacing or repeat spacing in a crystal
a_0	Bohr radius
$A(\mathbf{K})$	aperture function (includes physical aperture and aberration phase shift) $A(\mathbf{K}) = H(\mathbf{K})\exp(-i\chi(\mathbf{K}))$
A	atomic weight
A_n	numerical factor linking the aberration coefficient and probe angle to the diameter of the disc of least confusion
b	beam spreading (in cm)

Aberration-Corrected Analytical Transmission Electron Microscopy, First Edition.
Edited by Rik Brydson.
© 2011 John Wiley & Sons, Ltd. Published 2011 by John Wiley & Sons, Ltd.

B	brightness or magnetic field (vector, $\mathbf{B}$)
c	speed of light
C	concentration in weight percent
	(C_X, C_Y – concentrations in weight percent of elements X and Y etc.)
C_{atom}	concentration in atomic percent (or in absolute terms, atoms per unit volume)
C_c	chromatic aberration coefficient
C_{NS} or $C_{n,m}$	geometrical aberration coefficient (where the first subscript refers to the radial order of the ray deviation in real space and the second subscript the azimuthal symmetry)
C_s	spherical aberration coefficient equivalent to C_3 and $C_{3,0}$
d	probe diameter
d_c	diameter of the disc of least confusion (resulting from chromatic aberrations)
d_g	diameter of the disc of least confusion (resulting from geometric aberrations)
d_d	diameter of the diffraction disc
d_s	diameter of the Gaussian image of the source
d_{min}	minimum diameter of the probe at the optimum angle
D	dispersion of spectrometer
$D(\mathbf{K})$	detector function
$D_{ADF}(\mathbf{K_f})$	detector function (for annular dark-field detector)
$-e$	electron charge
E	energy loss
$\mathbf{E}$	electrostatic field vector
$E_c(\mathbf{K})$	partial temporal coherence envelope function
$E_s(\mathbf{K})$	partial spatial coherence envelope function
E_o	incident electron beam energy (in eV or keV)
E_p	bulk plasmon energy
ΔE	variation/instability in the electron energy (energy spread) or energy difference
δE	energy window
F	force (also a vector F)
$\mathbf{F}$	Fourier transform operator
Δf	defocus
f_{max}	maximum frequency response
f_{PCI}	phase contrast index function

g, h	reciprocal lattice vectors
h	Planck's constant
$H(\mathbf{K})$	Probe-limiting aperture, $H(\mathbf{K}) = 1$ inside and 0 outside aperture.
I_{probe}	probe current
$I_{\text{ADF}}(\mathbf{R}_{\text{p}})$	annular dark-field image intensity
$I_{\text{BF}}(\mathbf{R}_{\text{p}})$	bright-field image intensity
I_{xray}	denotes the characteristic X-ray peak intensities above the background
ΔI	instability in the objective lens power supply
J	current density
k	general electron wavevector ($\mathbf{k} = (\mathbf{K}, k_{\text{z}})$ ($|\mathbf{k}| = k = 2\pi/\lambda$)
$\mathbf{k}_{\text{p}}$	general incident electron wavevector
k_{XY}	*k-factor* or Cliff-Lorimer sensitivity factor
$\mathbf{K}_{\text{p}}$	transverse component of the incident wavevector
$\mathbf{K}_{\text{f}}$	transverse component of wavevector at the detector
L	camera length
M	magnification
n	number of electrons or the order of an aberration
n_{r}	refractive index
N	RMS noise or order of an aberration
N_{A}	Avogadro's number
N_{V}	Number of scatterers per unit volume
$O(\mathbf{R})$	object function
$p(\theta)$	scattering probability (as a function of angle)
q	general scattering vector
Q	image spatial frequency
r	Position vector $\mathbf{r} = (\mathbf{R}, z)$
r	distance (from the centre)
r_{total}	radius of the total disc of confusion
r_0	screening radius
R	bond length from an ionized atom
R	position in the sample (a two-dimensional vector perpendicular to z)
$\mathbf{R}_{\text{p}}$	illuminating probe position
S_{B}	signal from the surrounding background
S_{F}	signal from the feature of interest
ΔS	change in signal
t	thickness of specimen

$T(\mathbf{Q})$	optical transfer function for incoherent imaging
$V(\mathbf{R})$	specimen potential
$V_z(\mathbf{R})$	specimen potential projected along z
ΔV	instability in the accelerating voltage
v	velocity (in ms^{-1}) – also a vector v
w	complex scattering angle
W	error in the wavefront (measured as a distance)
W_B	intensity of the X-ray Bremsstrahlung over a specified energy range
z	distance along the optic axis
Δz	distance propagated along the z axis
Z	atomic number
Z_A	atomic number of atom A
Z_B	atomic number of atom B
α	semi-angle subtended by objective lens (numerical aperture), probe/beam convergence semi-angle (STEM/CTEM)
α_{opt}	optimum probe semi-angle, which can be a function of probe current
β	collector aperture semi angle (STEM), collection angle (EELS)
$\chi(\mathbf{K})$	aberration phase shift
δ	deviation (shift) of a ray from ideal Gaussian focus
Δ	focal spread distribution
γ	relativistic factor $= [1 - v^2/c^2]^{-1/2}$
λ	electron wavelength
Λ	mean free path
ρ	density (in g/cm^3)
$\phi(\mathbf{R})$	sample/specimen transmittance function
$\phi_g(\mathbf{K})$	amplitude of the transmission function for reciprocal space vector $\mathbf{g}$
σ	scattering cross-section/interaction constant between specimen potential and electron wave
Σ and σ	projected size and feature size in Ronchigram
τ	acquisition time
Ω	solid angle
$d\sigma/d\Omega$	differential scattering cross-section
ψ_o	incident wavefunction in CTEM
ψ_{exit}	exit surface wavefunction wave in CTEM
ψ_{image}	wavefunction in the image plane in CTEM

$\psi_p(\mathbf{R}-\mathbf{R}_0)$	complex amplitude of the STEM probe (probe function)
θ	angle (to optic axis)
φ	angle (azimuthal angle)
θ_B	Bragg angle
ζ	power of the atomic number for ADF imaging (also zeta factor in X-ray analysis)
μ/ρ	mass absorption coefficients for X-rays
Φ	magnetic scalar potential

Fourier Transforms

For a Free space plane wave, $\psi(\mathbf{r}) = \exp(i\mathbf{k} \cdot \mathbf{r})$

Fourier transform, $\tilde{\psi}(\mathbf{k}) = \dfrac{1}{\sqrt{2\pi}} \displaystyle\int \psi(\mathbf{r}) \exp(-i\mathbf{k} \cdot \mathbf{r}) d\mathbf{r}$

$\sim$ indicates the Fourier transform of a function

Inverse Fourier transform, $\psi(\mathbf{r}) = \dfrac{1}{\sqrt{2\pi}} \displaystyle\int \tilde{\psi}(\mathbf{k}) \exp(i\mathbf{k} \cdot \mathbf{r}) d\mathbf{k}$

$\mathbf{F}$ General Fourier transform operator (inverse $\mathbf{F}^{-1}$)

Fourier transformed quantity to be indicated by appropriate variable change in argument of function only

$\otimes$ denotes a convolution (defined below)

The convolution of f and g is written $f \otimes g$. It is defined as the integral of the product of the two functions after one is reversed and shifted. It is therefore a particular kind of integral transform,

$$(f \otimes g)(x) = \int_{-\infty}^{\infty} f(y)g(x-y)\,dy = \int_{-\infty}^{\infty} f(x-y)g(y)\,dy$$

Other Definitions

The phase object approximation is, $\phi(\mathbf{R}) = \exp(+i\sigma V(\mathbf{R}))$ where V is the electrostatic potential and is positive if attractive to electrons.

The propagation kernel is, $\exp(-i\pi\lambda\Delta z\mathbf{K}.\mathbf{K})$.

The lens transfer phase function is, $\exp(-i\chi(\mathbf{K}))$.

Aberrations: In this book we adopt the Krivanek-style notation for aberrations (see Appendix A). Here **positive defocus ($C_{1,0}$) corresponds to overfocus and a strengthening of a lens.**

ACRONYMS

ADC	analogue-to-digital converter
ADF	annular dark-field
BF	bright field
BFP	back focal plane
CCD	charge-coupled device
CTEM	conventional transmission electron microscope (or microscopy)
DAC	digital-to-analogue converter
DOS	density of states
DFT	density functional theory
DQE	detective quantum efficiency
DTSA	desk top spectrum analyser
EDX	energy-dispersive X-ray spectrometry
EELS	electron energy-loss spectrometer or spectrometry
EFTEM	energy-filtered transmission electron microscopy
ELNES	electron energy-loss near-edge structure
EXAFS	extended x-ray absorption fine structure
EXELFS	extended energy-loss fine structure
FEG	field-emission gun
HAADF	high-angle annular dark-field
HRTEM	high-resolution transmission electron microscope (or microscopy)
MAADF	medium-angle annular dark field
MDM	minimum detectable mass
MMF	minimum mass fraction
MSR	multiple scattering resonances
MTF	modulation transfer function
NIST	National Institute for Standards and Technology
OTF	optical transfer function
PCF	phase correlation function

PCF$_W$	phase-compensated phase correlation function
PCTF	phase contrast transfer function
RMS	root mean squared
SEM	scanning electron microscope (or microscopy)
SI	spectrum image (or imaging)
SNR	signal-to-noise ratio with subscripts i and o for input and output values
STEM	scanning transmission electron microscope (or microscopy)
TDS	thermal diffuse scattering
WPO	weak phase object
YAG	yttrium aluminium garnet

Index

References to tables are given in **bold** type. References to figures are given in *italic* type.

Aberration-Corrected Analytical Transmission Electron Microscopy, First Edition.
Edited by Rik Brydson.
© 2011 John Wiley & Sons, Ltd. Published 2011 by John Wiley & Sons, Ltd.

www.ingramcontent.com/pod-product-compliance
Lightning Source LLC
Chambersburg PA
CBHW072110250125
20788CB00003B/16